AF616621

Interchanges of Insects between Agricultural and Surrounding Landscapes

Interchanges of Insects between Agricultural and Surrounding Landscapes

Edited by

B. Ekbom
Swedish University of Agricultural Sciences,
Uppsala, Sweden

M.E. Irwin
University of Illinois,
Urbana, IL, U.S.A.

and

Y. Robert
INRA Lab Zoologie,
Le Rheu Cedex, France

SPRINGER-SCIENCE+BUSINESS MEDIA, B.V.

A C.I.P. Catalogue record for this book is available from the Library of Congress.

DOI 10.1007/978-94-017-1913-1

Printed on acid-free paper

Cover illustration:
Many pollinating wild bees, as for example members of the family *Megachilidae*, regularly change habitat between their nest site in hollow trees and open areas where they may pollinate agricultural plants. The cover photo shows a male *Anthidium manicatum*.
Photo: Mats W. Pettersson.

CONTENTS

CONTRIBUTORS

A. Andersen, Norwegian Crop Research Institute, Fellesbygget, 1432-Ås, Norway.

Jozef Banaszak, Department of Environmental Protection, The Pedagogical University Bydgoszcz, Chodkiewicza 51, Poland.

John Banks, Interdisciplinary Arts and Sciences, University of Washington, Tacoma, 1900 Commerce Street, Tacoma Washington 98402, USA. E-mail: banksj@u. washington.edu

Gary W. Barrett, Institute of Ecology, University of Georgia, Athens, Georgia, 30602-2202, USA. E-mail: gbarrett@sparrow.ecology.uga.edu

J. Baudry, Institut National de la Recherche Agronomique,SAD Armorique, 65 rue de Saint Brieuc, 35042 Rennes Cedex, France

Riccardo Bommarco, Swedish University of Agricultural Sciences, Department of Entomology, Box 7044, 750 07 Uppsala, Sweden E-mail: Riccardo.Bommarco @ wallace.teorekol.lu.se

Françoise Burel, Centre National de la Recherche Scientifique, URA 1853, Université de Rennes 1, Laboratoire d'Evoultion des Systèmes Naturels et Modifiés, Campus de Beaulieu, 35042 Rennes Cedex, France E-mail: Francoise.Burel@univ-rennes1.fr

Carlos Martin Cantarino, Dept. Ecologia, Universidad de Alicante, Apartado 99, 03080 Alicante, Spain.

Y. Delettre, Centre National de la Recherche Scientifique, URA 1853, Université de Rennes 1, Laboratoire d'Evoultion des Systèmes Naturels et Modifés, Campus de Beaulieu, 35042 Rennes Cedex, France.

Peter Dennis, The Macaulay Land Use Research Institute, Craigiebuckler, Aberdeen AB9 2QJ, UK E-mail: mi360@mluri.sari.ac.uk

Barbara Ekbom, Swedish University of Agricultural Sciences, Department of Entomology, Box 7044, 750 07 Uppsala, Sweden, E-mail: Barbara.Ekbom@entom.slu.se

G.L.A. Fry, Norwegian Institute for Nature and Cultural Research, P.O. Box 736, N-0105, Oslo, Norway

Carolina Godoy, Instituto Nacional de Biodiversidad, Santo Domingo de Heredia, Apartado Postal 22-3100, Heredia, Costa Rica, E-mail: cgodoy@rutela.inbio.ac.cr

Michael Irwin, Office of Agricultural Entomology, University of Illinois, 134 Envir. Ag. Sci., MC-637, 1101 W. Peabody Drive, Urbana, Illinois 61801, USA. E-mail: m-irwin2@uiuc.edu

Philippe Jeanneret, Eidgenössische Forschungsanstalt für Agrarökologie und Landbau Postfach, Reckenholzstrasse 191, CH-8046, Zürich, Switzerland. E-mail: Philippe.Jeanneret@fal.admin.ch

Gail E. Kampmeier, Center of Economic Entomology, Illinois Natural History Survey, 1101 W. Peabody, Urbana Illinois 61801, USA, E-mail: gkamp@ùiuc.edu

Andreas Kruess, FG Agrarökologie, Universität Waldweg 26, D-37073 Göttingen, Germany. E-mail: akruess@gwdg.de

Doug Landis, Department of Entomology and Center for Integrated Plant Systems, Michigan State University, E. Lansing, Michigan, 48824-1311, USA. E-mail: landisd@pilot.msu.edu

P.C. Marino, Department of Biology, University of Charleston, 66 George St. Charleston, South Carolina, USA; E-mail: marinop@cofc.edu.

N. Morvan, Centre National de la Recherche Scientifique, URA 1853, Université de Rennes 1, Laboratoire d'Evoultion des Systèmes Naturels et Modifiés, Campus de Beaulieu, 35042 Rennes Cedex, France.

Lowell R. Nault, Department of Entomology, OARDC, the Ohio State University, Wooster, Ohio 44691, USA, E-mail: lnault@osu.edu

Maurizio G. Paoletti, Department of Biology, Padova University, Via Trieste 75, 35100 Padova, Italy. E-mail: paoletti@civ.bio.unipd.it

S. Petit, Centre National de la Recherche Scientifique, URA 1853, Université de Rennes 1, Laboratoire d'Evoultion des Systèmes Naturels et Modifiés, Campus de Beaulieu, 35042 Rennes Cedex, France.

Yvon Robert, INRA Lab Zoologie BP 29, 35650 Le Rheu, France. E-mail: yrobert@rennes.inra.fr

Phil Taylor, Atlantic Cooperative Wildlife Ecology Research Network (ACWERN), Dept. of Biology, Acadia University, Wolfville, NS. B0P 1X0, Canada. E-mail: Philip.Taylor@acadiau.ca

Teja Tscharntke, FG Agrarökologie, Universität Waldweg 26, D-37073 Göttingen, Germany. E-mail: ttschar@gwdg.de

PREFACE

The chapters in this book were developed from some of the lectures presented at a symposium at the XX International Congress of Entomology held in Florence, Italy in August 1996. The purpose of the symposium was to discuss the impact of evolving modern agricultural landscapes on the insect species, of both economic and ecological importance, that utilize that habitat.

Agricultural policy, to some extent, influences the choices that farmers make and thereby the shape of the agricultural landscape. In order to move toward more sustainable agro-ecosystems future policy makers will have to consider the history of land use, consumer demands for both environmentally sound and affordable products, and the conservation of biological diversity. I would hope the information contained in this book will help stimulate discussion about the consequences of policy decisions on our agricultural landscapes and their insect inhabitants.

I thank all the speakers from the symposium and in particular those that have been able to contribute chapters to this book. There have been many delays, most due to circumstances beyond anyone's control. I would like to express my appreciation to Gloria Verhey and Patrick Dumont for taking care of the book in these final months.

CHAPTER 1

INTERCHANGES OF INSECTS BETWEEN AGRICULTURAL AND SURROUNDING LANDSCAPES

BARBARA EKBOM
Department of Entomology, Swedish University of Agricultural Sciences, Uppsala, Sweden

1.1 Introduction

Perhaps the single most important event in the history of humankind has been the domestication of the land. The rise and fall of civilizations has hinged on control and success of agricultural production. Feeding the populace has meant adding new land to the area devoted to crop production and technological advances have steered modes of production. Large units of production, fields and farms, worked by large machinery tailored to use of agrochemicals have dominated the development of agriculture in the last 50 years. No human activity has so changed the look of the landscape as has agriculture.

Increased agricultural production and the following manipulation of the landscape has not been without a price. Loss of natural habitats and associated species of plants and animals has not resulted only in diminishing biodiversity. It also threatens the very foundation of agriculture. Manipulations of the landscape based solely on human production goals are no longer possible because we cannot replace the environmental services on which the health of the ecosystem relies. Therefore we face the challenge of working with the environment using our knowledge of the characteristics of the system rather than imposing artificial structures which do not consider the fragile nature of the environment.

The problems we face are partly of our own creation because we have been unaware of or unable to address spatial and temporal aspects. Changes in the agricultural landscape have been primarily caused by economic and political considerations. In the future we must incorporate ecological thinking into production planning. This is not an easy task for two reasons. One is that it demands a relatively sophisticated integration of human and environmental needs and this is, of course, beyond the scope of this document. The second reason for difficulty is that we have only an incomplete understanding of the ecological processes involved given a landscape perspective. However, research on spatial and temporal aspects of ecological processes has made great strides and we can, more and more, not only make educated guesses but also provide data from relevant studies to answer the questions policy makers are asking.

The topic of spatial and temporal variation in relation to insect population dynamics is by no means new. Insect ecologists have long appreciated the fact that not only population size but also the stability of populations is affected by resource distribution and the ability of the insects to move between these resources (Gould & Skinner 1984).

B. Ekbom, M. Irwin and Y. Robert (eds.), Interchanges of Insects, 1-3

Although considerable progress has been made, especially through the efforts of conservation biologists, there is still a need for both development of theory and collection of empirical data. This book is an attempt to highlight empirical studies of insect-landscape interactions in the context of agroecosystems. The aim is to stimulate further study by presenting a variety of experimental techniques, statistical methods, and modeling approaches. Studies of processes at varying scales are complex and exchange of ideas on this topic is essential for designing future research.

This book focuses on one organism group in the agroecolandscape, insects. The focus is not arbitrarily selected. Insect pests are an age-old threat to agricultural production. Likewise cultivation of crops is dependent on insect pollinators, soil living insects that maintain soil fertility, and insect natural enemies important in suppressing pest populations. The insects we find in the agroecolandscape are, however, not restricted to the crop fields. The survival of most insect species is contingent on resources outside cultivated areas. Therefore management actions in fields will have far ranging consequences outside the field just as changes outside the field will profoundly affect insects within the field.

We have done little to encourage the beneficial insects in the agroecosystem. On the contrary we have contributed significantly to the well-being of the pests. We have developed crop varieties which are of high nutrient value, we have created vast areas with essential plant resources for herbivores, we have virtually removed the enemies of insect pests by indeterminate pesticide use and destruction of habitat. In time measured in terms of decades the wonders of chemical pest control were shown to have repercussions of frightening proportions. We now know that the use of insecticides in agricultural production is generally counterproductive from a long-term perspective. In the short-term, however, insecticides still are the most efficient means of pest control available. We must, therefore, present solutions carefully considering the consequences both in the long and short term and with the help of economists and sociologists find means of facilitating the change over to the landscape perspective.

Two major themes emerge on reading the chapters of this book: the structure of the landscape and the movement capacity of the insects. Translating individual behaviors on a small scale (both spatial and temporal) to the large scale is a challenge. Linking population dynamics to spatial and temporal habitat quality will be necessary to make predictions about the impact of landscape design. We will need to consider the landscape in terms of connectivity, configuration and composition. Connectivity has to do with ease of movement between habitats, high connectivity could potentially ameliorate effects of fragmentation. The composition of the landscape is the proportions of different habitat types available. Necessarily crop elements will represent a major proportion of the agricultural landscape's composition. For this reason the configuration of the landscape, which is the proximity of different habitats to each other, will become very important. Many organisms demand a variety of resource types and these resources must be within the movement range of the organisms.

Human perception and political considerations are a real part of the template on which we will design future agroecosystems. Two chapters (Burel - 2 and Paoletti, Cantarino - 3) discuss the changing landscape in relation to both scientific and human goals. Studies on different groups of insects with different ecological requirements will help us to focus on multiple spatial scales. We must appreciate the fact that changes in the landscape will have consequences at many levels from scales of a few square meters to hundreds of square kilometers. Management decisions can only be made by considering different organisms with a range of life histories and dispersal capacities.

Fragmentation of the habitat, caused primarily by loss of connectivity and dominance of crop fields in habitat composition, has been a major theme in conservation biology. This has great relevance in agricultural systems in light of our goal of preserving and enhancing populations of important natural enemies. Pests, predators, and parasitoids are considered in a landscape context in three chapters (Kruess & Tscharntke - 4, Barrett - 5, and Jeanneret - 6). Two other chapters (Taylor - 7 and Banaszak - 8) explore the impact of fragmentation from a more general biodiversity point of view. Taken together these chapters focus on fragmentation and do so using widely disparate systems and insects with very different lifetime spatial ranges. A variety of experimental methods and modeling approaches are presented.

The last half of the book (Irwin, Nault, Godoy & Kampmeier - 9, Bommarco & Ekbom - 10, Landis & Marino - 11, Dennis, Fry & Andersen - 12, Banks - 13) is made up of studies of specific systems and all focus on pest management enhancement by cultural methods. Dispersal ability of both natural enemies and pests are discussed as well as requirements for habitats of high quality for natural enemy population development.

We will never be able to study all the insects in the agroecosystem, parameterize their movement behavior, and ascertain their habitat requirements. We can, however, address ecological questions using groups of organisms with similar biologies, dispersal modes, and resource requirements (Duelli & Obrist 1995). We will also need clear objectives. Do we want a simple increase in biodiversity or an increase in abundance of specific organisms? Is this a paradox or can these conflicting goals be resolved? Ecologists can clearly help to refine objectives by describing the consequences of different management activities, but final decisions are made by the community.

To say that we should make decisions based on unproved theories and haphazard knowledge because we must act quickly is to play the fool. No policy decision should every be hard and fast. Respect for the limitations of our knowledge should be factored into management packages by continuing support for basic research and monitoring of effects of changing land use against clearly defined goals. There should be time to do the studies and test the theories - at the necessary scale. Conclusions must be robust and directions given for fine-tuning of landscape designs in accordance with the climatic/physical environment as well as the culture and traditions of the society.

1.2 References

Duelli, P. & Obrist, M. 1995. Comparing surface activity and flight of predatory arthropods in a 5 km transect. In: Toft, S. & Riedel, W. (Eds.), *Arthropod natural enemies in arable land, I. Density, spatial heterogeneity and dispersal. Acta Jutlandica* **70**:2, pp. 283-293.

Gould, F. & Stinner, R.E. 1984. Insects in Heterogeneous Habitats. In: Huffaker, C.B. & Rabb, R.L (Eds.) *Ecological Entomology*. Wiley Interscience, New York, pp. 427-449.

CHAPTER 2

RELATING INSECT MOVEMENTS TO FARMING SYSTEMS IN DYNAMIC LANDSCAPES

F. BUREL
Centre National de la Recherche Scientifique, UMR Eco-Bio,
Université de Rennes 1, Rennes Cedex, France

J. BAUDRY
Institut National de la Recherche Agronomique, SAD Armorique,
Rennes Cedex, France

Y. DELETTRE, S. PETIT, *and* N. MORVAN
Centre National de la Recherche Scientifique, UMR Eco-Bio,
Université de Rennes 1, Rennes Cedex, France

2.1 Introduction

During the last decade the conservation of biodiversity in agricultural landscapes has been greatly emphasized. Intensification of agricultural production and practices, as well as land abandonment, have been considered as major threats (Solbrig, 1991), by making drastic changes in landscape structure and composition. The agricultural landscape is a shifting mosaic of crops, pastures, fallow lands and woody areas. Landscape elements exhibit their own disturbance regime, which depends on the practices used by farmers and thus interact with insects at several spatio-temporal scales. Changes at regional and long-term scales are the most documented and the most predictive (Baker, 1989; Rackham, 1986; Odum & Turner, 1990; Meeus, 1995; Crumley & Marquardt, 1987). A recent trend in western Europe is a decrease of the area covered by non-cultivated elements such as hedgerows, woodlots, heathlands, within intensive agricultural landscapes (Agger & Brandt, 1988; Bunce & Hallam, 1993; Burel & Baudry, 1990; Morant *et al.*, 1995). In the mean time large tracks of farmland are abandoned. Thus the contrast between different regions increases. The changes within rural landscapes result in an increase in fragmentation of many elements and affect insect populations by reducing available habitats or seasonal refuges for many species. At a finer scale, farmers make decisions on crop succession in their farming system and on management practices for field boundaries and non-productive areas; this creates a changing landscape mosaic. Changes in agricultural landscapes can only be predicted through the knowledge of how farmers will change the land use pattern under different circumstances. Two levels must be considered: 1) changes in the type of production (*e.g.* from dairy production to cereals) and 2) changes in the techniques of production (*e.g.* feeding dairy cows with hay or

B. Ekbom, M. Irwin and Y. Robert (eds.), Interchanges of Insects, 5-32

maize silage). If, at a broad regional scale, relationships between the type of farming systems and landscape can be established, it becomes very fuzzy at the landscape scale, relating to, the scales of individual and population dispersal. Deffontaines *et al.* (1995) provide examples of two farms, in the Pays d'Auge, where the most productive one has more grassland and less annual forage crops. In all cases, the within farm land use diversity is striking, it results from both physical and spatial constraints and the requirements of the system of production (*e.g.* winter/summer forage). Many factors, out of the agricultural sector, also affect land use and landscape patterns, as in multi-job farms (Laurent *et al.*, 1994). Trajectories of households as well as changes within farm systems are barely related at the individual farm scale. The major consequence from a landscape ecological point of view is that landscape changes cannot be derived from current landscape patterns (Burel & Baudry, 1990). More specifically, abandonment or dereliction of grassland is a stochastic process at the landscape scale, although deterministic at the farm scale.

In heterogeneous landscapes the interaction between mosaic structure and movement patterns determines the distribution of individuals in space, and ultimately how spatial patchiness affects population and community dynamics (Wiens *et al.*, 1993). Because of the strong influence of Island Biogeography Theory (MacArthur & Wilson, 1967) the first research carried out on animals movements in agricultural landscapes focused on species in woodlots and forests or woody corridors (Verboom & van Apeldoorn, 1990; Bennett, 1990). In these cases farmland is considered as an hostile or neutral matrix. This approach has been shown to be insufficient (Merriam, 1988), even for the study of "forest species", as they can move through or use adjacent fields for part of the year. Heterogeneity of the landscape has to be considered, in relation to species requirements.

Movements may be induced by foraging behavior, mating, metapopulation dynamics, species interactions, and so on. The different life processes and their associated movement behaviors occur at different spatial scales, from foraging in a local patch to migratory movements on regional or continental scales. Whatever the type of movement, it will depend on the interaction between the landscape structure and the movement patterns (Wiens *et al.*, 1997). These interactions determine the resistance or facilitation induced by the different landscape elements all along the pathway followed by a moving individual. They are species dependent so there is no unequivocal response to a given landscape structure (Gustafson & Gardner, 1996). Recent researches on this topic led to the creation of several models using artificial organisms moving within simulated or real landscapes (Wiens & Milne, 1989, With *et al.*, 1997). Only a few examples are available using real organisms in real landscapes to validate models (Schippers *et al.*, 1996) and empirical data are scarce. Most of the recent results deal with dispersal movements in a metapopulation context. In this context studied species are restricted to one habitat type and emigration or immigration rates are related to patch size, edge permeability and movement corridors (Stamps *et al.*, 1987; Fahrig & Merriam, 1985). Nevertheless multi-habitat species that are not restricted to one vegetation type but depend rather on the landscape mosaic are quite common. All the different landscape elements linked by a single biological process (here individual movement) form a functional unit (Merriam, 1984; Morvan *et al.*, 1994). Survival of an individual or a population within a given landscape depends on the integrity of the functional units, measured as connectivity and heterogeneity of the landscape mosaic.

In this paper, we focus on the consequences of changes in agricultural landscapes on insect movements; such changes on mosaic heterogeneity in relation to crop succession or the quality of uncultivated habitats in relation to management techniques and periodicity. Our approach is summarized in Figure 1.

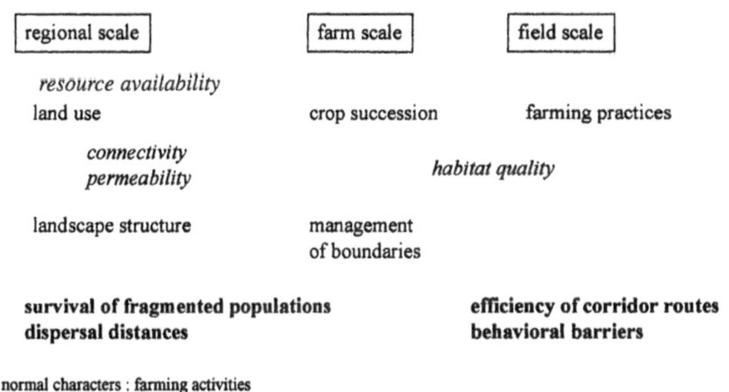

Figure 1. Conceptual framework of the research.

Farming activities and insect movements interact at several spatio-temporal scales, depending both on landscape structure and species biological features. As scale is central in our approaches, hierarchy theory provides a convenient frame to analyze spatial patterns dynamics (Allen & Hoekstra, 1992). The space can be divided into levels of organization that are characterized by their own dynamics. According to the theory, the structures at the different levels of organization are controlled by different factors. It also states that higher levels of organization constitute a context of constraints for the functioning of lower levels.

In this paper we present some results from multi-disciplinary research on hedgerow network landscapes, initiated in Brittany, France, since 1986. Several landscape units have been investigated to illustrate relationships between landscape dynamics and insect populations. In contrasted landscape units, controlling factors of landscape dynamics are identified at various social and spatial levels. Their effects have been studied on the movements of multi-habitat species (Diptera; Empididae and Chironomidae), and on stenotopic (*e.g. Abax parallelepipedus,* Coleoptera, Carabidae). From the results some guidelines may be suggested that will facilitate or inhibit movements of beneficial, pest, or endangered species. Empirical data are used to build spatially explicit models.

2.2 The dynamics of agricultural landscapes

2.2.1 HISTORY OF HEDGEROW NETWORKS IN BRITTANY (FRANCE)

Rural landscapes, in western France, are characterized by the presence of hedgerows inter-connected into a network and connected to uncultivated areas such as woodlots, heathlands and old fields. They have been dominated by human influences for centuries. Their current form results from environmental constraints, from agricultural techniques and practices, and from farmers' attitudes, as well as from other socio-economic and political factors (Burel & Baudry, 1995a).

There is evidence that the first hedges date from the Roman and even pre-Roman times (Rackham, 1986; Morgan Evans, 1992). Extensive clearing of the primitive forest during the XIth and XIIth centuries may be related to technical progress such as the wheel plough. Until the XVIIIth century moorlands and cultivated areas are mixed with grazed areas. Within cultivated areas hedgerows delimit fields, preventing cattle from feeding on crops, and limiting conflicts among shepherds (Meyer, 1972). Construction of hedgerows has been a slow process, with periods of construction and periods of abandonment following the rises and falls of human populations.

The eighteenth and nineteenth centuries have been enclosure periods in the France and Great Britain as new techniques permitted an increase in the surface of cultivated land.

Since World War II there has been a drastic decline worldwide of the number and length of hedgerows in the traditional hedgerow network, or "bocage" landscapes. Fields have been enlarged to facilitate the use of large machinery leading to the removal of shrubs and trees that limited previously small fields. In Europe remaining large and tall hedgerows are mostly located along pastures, whereas arable plots are usually bordered by thin hedges when they are still present (Barr, 1993; Hegarthy *et al.*, 1994). Indeed, on arable farms, hedgerows have lost their previous purpose of shelter and fence for cattle (Boatman, 1994).

Those changes can be illustrated by the case of the municipality of Lalleu, Brittany, studied by Burel & Baudry (1990) before a reallotment program and investigated afterward (Table 1). We characterize the hedgerow nework using three parameters: hedgerow length, connectedness (the number of connections among hedgerows) and the number of no-connections (number of ends of hedgerows not connected to any other). The decrease of hedgerow length and connectedness has been constant since 1952 and has been amplified by the reallotment program which has resulted in a massive removal of hedgerows. The number of no-connections, hence isolated hedgerows, has increased.

Table 1. Changes in the hedgerow network in the municipality of Lalleu (Brittany)

		Hedgerow (m.ha-1)	Connectedness per ha	# no-connections
1952		225	10	280
1974		185	6	510
1988	Realottment	152	4.5	590
1989	program	60	1	1480

According to their age, their location and their use by farmers, the vegetation of hedgerows varies greatly (Pollard *et al.*, 1974, INRA *et al.*, 1976), as well as the structure and extent of their basal area. In past centuries in Europe, most field boundaries were used for drainage or irrigation purposes, and thus bordered by a ditch. The earth dug out for the construction of the ditch was piled up to form a bank on which the vegetation developed (Dowdeswell, 1987). The complexity of this structure depends on the local traditions, the stoniness of the soil, and the type of agriculture. It plays a major role on ecological processes driven by hedgerows and the networks they form. In most cases recent windbreaks are planted at the soil level which diminishes considerably the diversity of the herbaceous vegetation, and the impact of the linear element for water and erosion control.

Technology, ownership patterns, culture and religion are reflected in the different management techniques found in the "bocages" (Bannister & Watt, 1994; Watt & Buckley, 1994). Trees used to be coppiced, pollarded or grown as timber, depending on the target species, and on their local uses (Luginbühl, 1995). For example in eastern Brittany oaks used to be pollarded, as their trunk belonged to the owner of the land whereas the branches belonged to the tenant. Tenants were allowed to cut branches every 9 to 12 years according to the length of their tenure. Several types of tree management may coexist on a single hedgerow, creating a complex structure that offers a diversity of microhabitats for animals. Hedgerows are trimmed or branches cut periodically. This creates, at the landscape level, a shifting network of successional stages, as not all hedgerows are managed at the same time. This spatio-temporal heterogeneity is followed by insects, as for example *Miridae* (Heteroptera) assemblages, that shift from a dominance of aphid predators just after the branch cutting, toward a dominance of phytophagous species when branches are overgrown (Ehanno, 1988).

2.2.2 CURRENT AGRICULTURAL LAND COVER AND LAND USE DYNAMICS

Current status of the landscape structure is inherited from past history, but farming activities play a major role in the yearly dynamics and are responsible for most of the changes encountered today. In this presentation of agricultural landscape dynamics, as a framework for species dispersal, we concentrate on two goals: 1) to explore landscape changes at various scales and 2) to assess the possibilities to elucidate mechanisms of change in order to simulate landscape dynamics and its consequences for insect populations. We intend to expand the conceptual framework that Burel & Baudry (1995b) have developed regarding landscape as a medium between farming activities and species dispersal. Examples are drawn from our current studies in Brittany, France.

The conceptual framework

The dynamics of hedgerow networks is highly linked to the dynamics of farming systems and land use/cover changes. We explore those changes in this section. We use the distinction between land cover and land use in the sense of Meyer & Turner II (1994). Land cover is "the quantity and type of surface vegetation, water ...", land use is "the human employment of the land". The important point is that differences in land covers are conspicuous (a forest, a cereal field), while differences in land uses may be difficult to notice (amount of pesticide, type of plowing). We know land covers from aerial photographs, censuses, land uses by short time scale surveys or land users' interviews. From a given ecological process (foraging, mating), the differences between two types of land covers may be less important than the differences of practices between two farmers managing the same type of crop; conventional versus organic farming is but one example.

At coarser spatial and time scales, we use land covers and land uses at finer scales. Land cover types control the presence of species over a region, farming practices drive individual behavior.

When considering the dynamics of agricultural landscapes two views are possible. On one side, patterns at a regional level result from the aggregation of patterns at field and farm scale; farm statistics being made from the survey of each and every farm or from sampling. On the other side, the regional level is also a level of political decisions (Laurent & Bowler, 1997) and usually regions are defined according to specific geomorphological, climatic and historical features that set a context for the evolution of farming systems. Thus, the region can be considered as an autonomous system that only permits certain types of farming systems to be present at finer scales.

We can hypothesize that the changes we perceive at different scales affect different ecological processes, therefore, it is worth looking at those scales from region to field.

The results

At the regional scale (several thousand square kilometers), knowledge of changes in types of land cover over decades are provided by agricultural censuses. Farming involves first clearing of the land for grazing and plowing, other important changes are shifts in crops, some are no longer cultivated, while others are adopted. These trends can be exemplified by the decrease of moorland in Brittany for the last century and the recent increase of maize as the fodder crop at the expense of leguminosae (clover and alfalfa) (Fig. 2). As machines to grow and conveniently harvest maize became available, the crop could extend, which, in turn, required the development of new breeds, adapted to cooler climatic conditions.

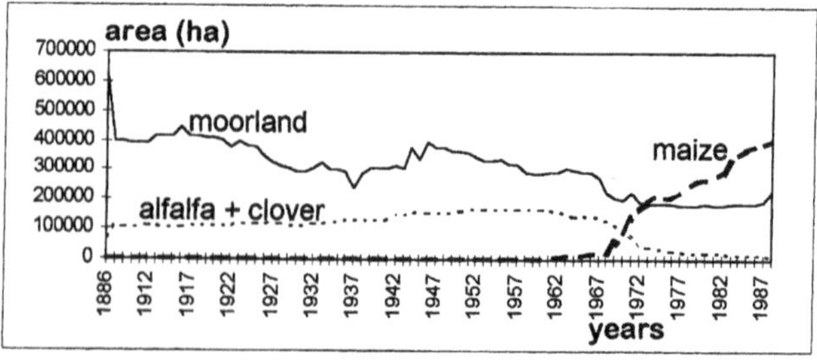

Figure 2. Changes in crop production in Britanny, France, from 1880 to 1990.

Another frequent type of change is the shift between cropland and permanent grassland as exemplified by the case of Lower-Normandy (Fig. 3). Those two types of land cover differ by the frequency of soil disturbance, while differences between alfalfa and maize are related to food source. Species with different life history traits will be affected by either type of change.

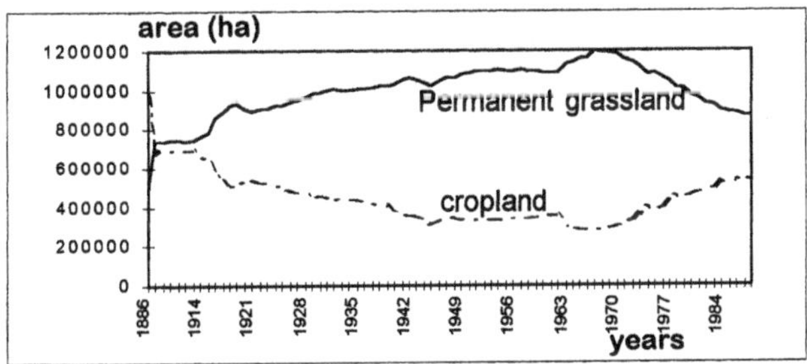

Figure 3. Shift between cropland and permanent grassland in Normandy, France during the 20th century.

These examples show that the paces of changes can be quite different, the shift from leguminosae to maize has been very rapid. The decrease of moorland, by converting them to forest or to farmland is much slower and periods of extension may occur as during the wars. As farming activities are under the constraints of national and international policies, the extension of maize may be reversed. Subsidies for its cultivation are currently high, but as it is an important cause of nitrogen leaching and water pollution, maize may be banned and replaced by alfalfa. Eventually we may obtain a pattern similar to the shift between grassland and cropland in Normandy. In any case the highly dynamic pattern of land cover at the regional level is the first fact to be taken into account. It may prevent slow colonizers from being present everywhere in a given crop. Baudry (1992), analyzing the perception of changes at various spatial and time scales, shows that the coarser the spatial/time scale, the lower the changes. The scale dependence of the rates of changes must be taken into account when making comparisons across regions or periods. Dissimilar patterns may be related to different choices of scales and lead to wrong conclusions (Allen & Hoekstra, 1992).

At the landscape level (a few tens to a few hundreds of hectares), several features are important: the proportion of the different land cover, their arrangement in space, the year to year crop succession and a set of biological (plant development) and technical (farming practices) events occurring within a single year.

The proportion of the different land covers and crops is a function of the regional type of production (cereals, forage ...). Nevertheless, differences in the technical culture of regions lead to differences in land uses for a similar final product as shown for the European Union by Baudry *et al.* (1997).

Crop succession from year to year has been a basis of good agronomical practices for several centuries. Except for permanent grassland and woodland, the cultivation of the same crop for several years in the same field can lead to disease problems or loss of nutrients. Therefore, the exact placement of a given plant crop will change as well as its relationship with adjacent crops or field boundaries. In bocage landscapes, harvest of fire wood every six to ten years also creates a pattern of changes.

Within a given year, reasons for changes are many. The first set of causes is vegetation growth, for example, maize (*Zea mais*) offers a bare ground up to mid-May and a dense vegetation cover from July to September, while fields of other cereals are devoid of tall vegetation. Flowering is an important period in plant life cycle that interacts with many insect life histories (pollination, seed production). Farming practices are the second set of causes. Farmers have to prepare the soil, harvest, spray ..., creating at every step a new environment. In the next section, we will see that vegetation growth and farming practices have strong interactions.

The study sites

Our analysis of the effects of farming systems on landscapes are conducted at three sites where the Diptera are also investigated. Three sites were chosen in northern Brittany, south of the Mont Saint Michel Bay, from aerial photographs. Their area ranged from 500 to 700 ha. They were distributed along a gradient of landscape openness from a dense hedgerow network (site A) to a sparse one resulting from a reallotment programme (site C), site B being in an intermediate position. In this region, the agriculture is oriented towards milk production and the farming system is identical in the three sites. Permanent grasslands are mixed within a mosaic of different cropfields. Both hedgerow and crop field attributes were surveyed and entered in a spatially explicit database related to Geographical Information Systems (Arc-Info™ and IDRISI™) which provide accurate maps and different metrics of landscape structure such as heterogeneity, fragmentation, connectedness and so on (Table 2).

Table 2. Some general parameters of the three study sites (after Morvan, 1996)

	Site A Broualan	Site B Vieux Viel	Site C Pleine Fougères
Total area (ha)	531	637	471
Average patch size (ha)	1.33	1.53	2.64
Average hedge length (m/ha)	204	193	76
Permanent grassland (%)	39.7	29.7	18.0
Temporary grassland (%)	13.4	13.0	9.3
Maize (%)	20.1	18.7	39.8
Cereal (%)	12.5	24.3	23.0
Wood (%)	5.8	3.0	3.1
Fallow (%)	0.8	1.6	0.5
Water bodies (%)	0.6	0.6	0.5
Roads and houses (%)	6.8	9.1	5.7
Global heterogeneity (Shannon's diversity H', mosaic + hedges)	1.80	1.65	1.61

From a general point of view, the global heterogeneity decreases from site A to site C. This fact is not related to a decrease in the diversity of landscape features but, rather, to a different mosaic structure. Furthermore, site C exhibits a higher average patch size and a much lower hedgerow length, owing to land reallotment.

2.2.3 FACTORS DRIVING THE DYNAMICS

Because all these changes have consequences for habitat, food, reproduction site availability, it is necessary to analyze their causes within a farming system context. This will enable us to assess the consequences of changes within farming systems on species dispersal, or to incorporate practices enhancing or restricting movement. The idea, in itself, is not new and is at the foundation of practices aiming at reducing pest damages in integrated agriculture (Altieri, 1983, Glen *et al.*, 1995).

Field investigations and interviews of farmers in different landscapes of western France, where dairy farming is the dominant system, permit an overview of driving factors in three areas:

1) the spatial distribution of land utilization: how differentiations among farming systems and within farm technical system make a land mosaic within a given ownership structure (scattered *vs.* grouped fields).
2) the relationship between hedgerow structure and land use, field margin vegetation: hedgerows along permanent grassland have a denser tree and shrub cover than those along crop fields. The herbaceous layer structure (field margin) is also influenced by land utilization: if the field is grazed by cattle, the margin will be different than if it is never grazed but only mown or sprayed with herbicides.
3) these field margin management practices are strongly related to the type of farmer, depending on both their perception of hedgerows, the availability of labor and cash to buy herbicides.

Here we define a field as the spatial unit where, during a cropping season homogeneous practices are carried out. It is, thus, defined by land users (farmers) as a management unit. The field, its position in the landscape and its integration in a farm is a key element for understanding landscape patterns and dynamics. It is the very place where decisions are made in terms of management as well as in terms of landscape design (field enlargement, hedgerow planting).

Land utilization

At a global scale, land cover and use are a function of the type of production in the different farms (*e.g.* cereals as cash crops, maize and grassland as forage). Within an average dairy farm, the different land uses are roughly distributed as follow, starting from the farm buildings: long term grassland to graze dairy cows (cows cannot walk more than one kilometer). The succession includes wheat or maize for silage every five to six years, but most of maize is further away and cereals still further. Distant, small or hydromorphic fields are utilized as permanent grassland for heifers. Thus, there is a basic concentric distribution model more or less disturbed by stream corridors or roads. In fact farmers avoid having their cows cross a road to walk from field to barn for milking twice a day.

At the landscape scale, this basic module will show up if farm territories are in a single or a few groups of fields. If, as in many regions, the fields of the different farms are mixed because of successive divisions among heirs, the land use mosaic may not exhibit any clear spatial organization. It will be very heterogeneous and the spatial relationships among the different types of patches unpredictable. The crop succession in most parts of the landscapes will be unpredictable as well. Exceptions are permanent cover such as grassland or woodland.

Hedgerow and hedgerow network structure

The survey of hedgerow structure in our three study sites shows a relationship between this structure and the adjacent land cover (Le Cœur *et al.*, 1997). The denser hedgerows are along woodland and permanent grassland. Around cropland hedgerows have an average tree cover of less than 50% and usually have gaps. Hedgerows between grassland and cropland have an intermediate structure. Roads are bordered by treeless hedgerows, only a scattered cover of shrubs is present. Thus, the fields with the highest rate of disturbance (plowing, spraying for crops) are related to open hedgerows. These fields are also the larger fields.

From this case study, Thenail & Baudry (1996) propose a series of driving factors of changes at various levels. They stress the importance of the technical organization of the production within farm (machines, land, labor ...), but demonstrate that external factors can also be very powerful (policies, land reallotment ...). Thus, at any scale, *i.e.* for all ecological processes, there are possibilities to predict dynamics resulting from the functioning of agricultural systems that will influence those processes. This should enable us to build models. In landscape ecology, the growth of empirical evidence and concepts has been rapid this last decade, but in agronomy and farming system research, research dealing with land utilization patterns at a spatial level higher than a farm is still in its infancy. An important research question is whether new concepts and theory on agricultural systems to understand spatial patterns can be developed for their own sake or with the goal of making connections with ecological processes?

An example

These dynamics are illustrated in Figures 4 and 5. Figure 4a shows the change in land cover over a three year period. Not only is there a shift between maize and other crops (mainly winter wheat), but in 1996 the surface of maize increases strongly to become the dominant land cover. Figure 4b illustrates some consequences on the state of the land within a year. Plowing for maize is done in spring, while it is done in autumn for winter cereals. Plowing can be a double source of disturbance: i) turning up the soil and exposing larvae to the surface and ii) destroying the herb layer with associated seeds which is a source of food. Depending upon the crop these disturbances occur at different periods. The period where the crops are present also changes, so their function as shelter. Maize being present during the summer, it is very difficult for farmers to access the field margins to mow them, as they do after wheat harvest. In the case of maize they spray herbicides in June or early July before the full development of the plants (Fig. 5).

Therefore the decision made by a farmer to grow a given crop has consequences beyond the presence of such or such a species. It affects the state of the landscape between crops and even the disturbance regime on field margins.

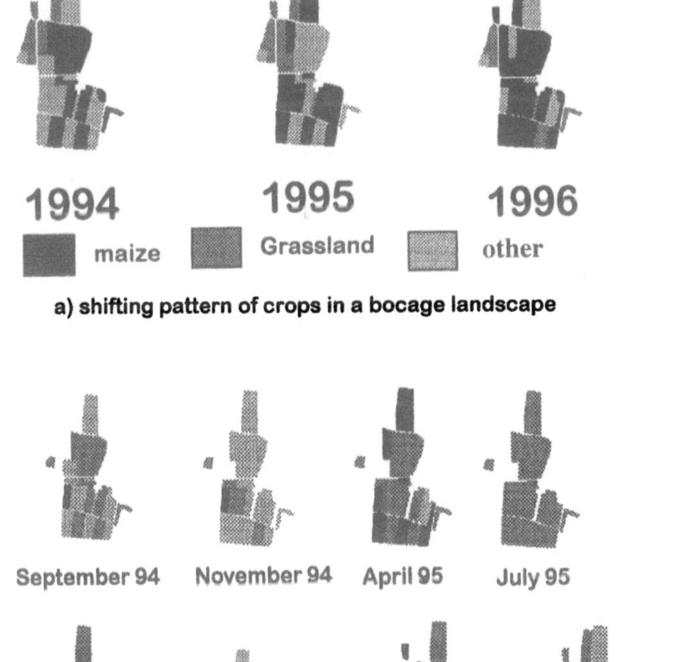

Figure 4. Maps of land cover change in part of a study site. a) at yearly scale; b) at monthly scale.

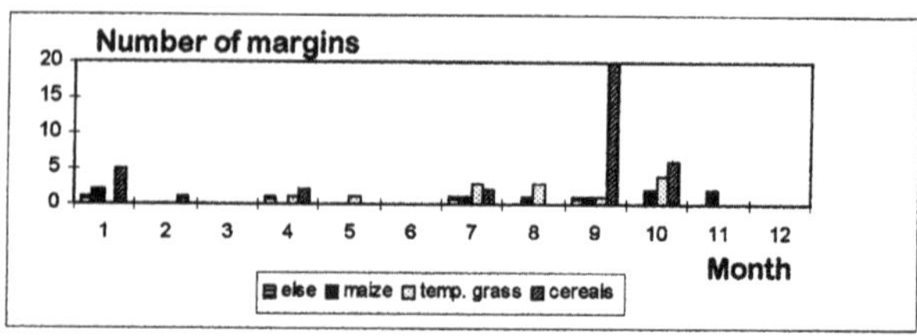

a) mowing

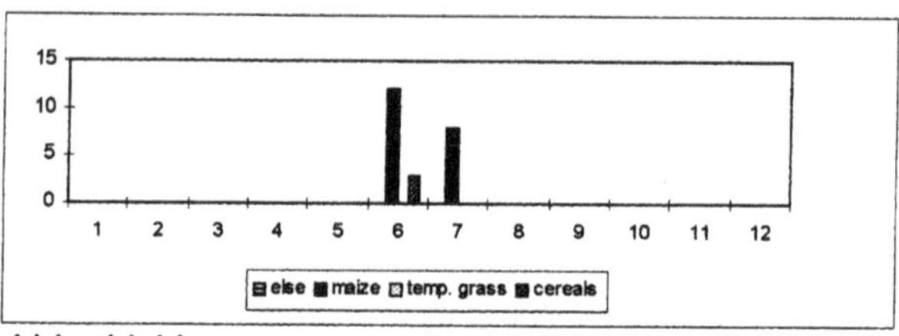

b) herbicides

Figure 5. Management practices on field margins; a) mowing, b) herbicide use.

2.3 Insect Movements in Agricultural Landscapes

In agricultural landscapes, Duelli *et al.*(1990) distinguished insects according to their temporal and spatial distribution. Four main classes were identified: (1) stenotopic species spending their entire life cycle within one type of habitat, (2) species with a stronghold in or around natural or semi-natural habitats, (3) fliers using field margins in or outside the study area for hibernation, (4) nomadic or migrating species evenly spread over an entire study area. According to Merriam (1984), a fifth class must be added for species which need different landscape units to complete their whole life cycle (landscape functional unit).

Depending on the type (walk, passive or active flight) and intensity of movement (mean distance covered per time unit), species will interact with the spatio-temporal structure of the landscape

* movement corridors will be used by species for finding cover, food, favorable physical conditions. For example hedgerows or heathland road verges have proved to be efficient corridors for the dispersal of some forest insects (Burel, 1989, Dover, 1991) or heathland species using road verges as habitats and corridors linking heathland patches (Vermeulen, 1995).

* movement between fields and boundaries is affected by the nature of the crop, the size of the field and the structure or nature of the boundary (Thomas *et al.*, 1992; Wratten & Van Emden, 1995)

Being affected either by corridor networks or by heterogeneity of the mosaic movements depend on farm activity at various spatio-temporal scales. In the following sections we present some results on the effects of mosaics on species using different landscape elements such as Diptera (Chironomidae and Empididae), and corridors for some stenotopic species such as a Coleoptera:Carabidae, *Abax parallelepipedus* (Piller & Mitterpacher, 1783).

2.3.1 SPECIES USING HETEROGENEITY OF THE CROP MOSAIC

The Diptera, Chironomidae

The Chironomidae are frequently found among the numerous dipteran families that fly in agricultural landscapes. Although several species originate in different soil types including crop fields (Strenzke, 1950; Sendstad *et al.*, 1977; Seddon, 1986; Delettre, 1984; Hudson, 1987; Delettre & Lagerlöf, 1992), most of them emerge mainly from water bodies like temporary or permanent ponds, brooks, rivers and lakes (Armitage *et al.*, 1995). While numerous papers have been published on their ecology in aquatic ecosystems (Fittkau *et al.*, 1976; Hoffrichter, 1981), the influence of landscape structure and heterogeneity on the dispersal of aquatic species has not been investigated previously. Landscape structure must, however, be considered to explain the dispersal of flying individuals between isolated water bodies or subdivided populations.

Both terrestrial and aquatic species were caught during a two-year study in the three study sites of Northern Brittany. Data processing is still in progress but results from early spring emergence at site B (Vieux Viel) already suggest a role of landscape structure on the spatial distribution of the adult stage of aquatic Chironomidae. Insects were sampled using 52 yellow traps set in pairs on the soil surface at the bottom of 26 hedges of different types over the whole area. Traps were opened three days each week. As females are not described in most species, only males were identified. Relationships between their spatial distribution and landscape attributes were analyzed using Canonical Correspondence Analysis (Ter Braak, 1987), Correspondence Analysis and Hierarchical Classification (Legendre & Legendre, 1984).

From the different analyses, it was clear that the distance between traps and water bodies was the main factor explaining the spatial distribution of adult midges at the studied site (logarithmic regression, $r = 0.629$, $n = 52$, $P < 0.01$): the more distant the traps, the lower the abundance of flying Chironomidae. This first result shows that adult non-biting midges originating in water bodies exhibit a stronghold around their emergence sites although they can be found flying over the entire study area.

However, the distance to the nearest water body was not the sole parameter explaining chironomid distribution at site B. The quality of hedges is also a significant factor. An hedgerow typology was computed from 10 descriptive attributes (hedgerow permeability, canopy width and height, cover of three different plant strata, crop field use on each side of the hedge, width of the uncultivated field margin, total width of the -bank+ ditc-strip), which resulted in six different hedgerow classes. This typology was independent from the distance to water for five classes: no significant difference was found between average distances to water bodies. Only the treeless hedges (class 6) were located at a significantly longer distance from water bodies.

When the average midge abundance per trap was plotted against the average number of species in each hedgerow class (Fig. 6), a significant linear relationship occurred (r = 0.975, n = 5, P0.05). The number of species and individuals is higher in hedges with a high tree height and a well-developed shrub cover (class 1) while these numbers are lower in hedgerows with very narrow uncultivated field margins and almost treeless (class 6). Other classes were in an intermediate position.

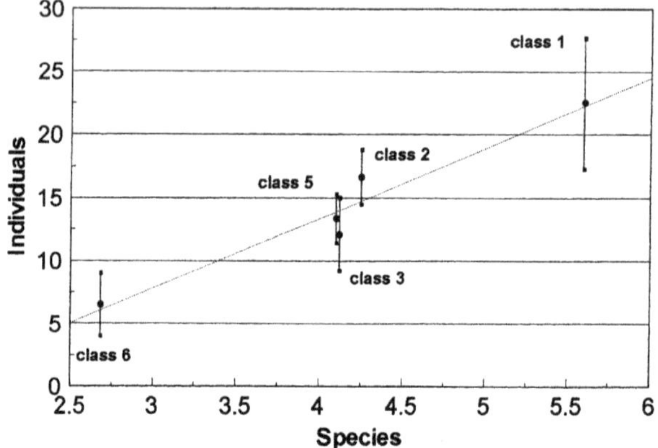

Figure 6. Average number of individuals (± SD) and species per trap in the different classes of the hedgerow typology (class 4 is absent from site B).

This result suggests that the hedgerow quality is an important factor that influences the number of adult midges at rest in the foliage during a part of the day (adult midges only swarm at dawn and dusk and spend the rest of the day at rest, without feeding).

Hedgerow quality directly depends on their management by farmers (Pollard, 1974) but also on land use in adjacent plots (Le Cœur, 1997). Thus, agricultural practices are likely to influence midge distribution through hedgerow management.

Although further research is needed to understand more thoroughly these relationships, it seems that landscape structure and heterogeneity interact both at a landscape scale (distance) and at a local scale (hedgerow quality) with the spatial distribution of adult midges, a fact which is likely to influence exchanges between chironomid populations located in water bodies scattered throughout the landscape.

Furthermore, the current results show that species life-history traits as well as seasonal dynamics must be considered to characterize a given landscape and to approach more closely real processes in the field. For instance, the hedgerow typology mentioned above is independent from the distance to brooks for five of its six classes whereas Le Cœur (1997) showed that dense hedgerows were more numerous near permanent grasslands, which are preferentially located near brooks. In our case, this independence comes from the fact that all the potential sources of aquatic chironomids were considered, including not only brooks but also ditches flooded in early spring, where numerous larvae were observed. Probably, it is not necessary to take ditches into account to characterize this landscape in a general manner but, for those species at this particular moment (early spring) flooded ditches are important. Some time later (*i.e.* during summer), ditches can be neglected because all of them are dry and no longer produce any adult chironomid.

The next step of this research is to test how this relationship between chironomid distribution and distance to water bodies evolves in different landscapes. For this purpose, data from the three study sites are concurrently considered. If midge dispersal is limited by species-specific flying capabilities, the same result is to be expected whatever the landscape openness. On the contrary, if landscape structure modifies species dispersal, a longer average dispersal distance should be observed in more open landscapes. Hedgerow number and quality will be considered as well. In a second step, an attempt will be made to relate the distribution of several species which actively prey on chironomids (like Empididae or insectivorous birds) and the observed distribution of adult midges. If some kind of spatially explicit relationship occurs, this work will evolve towards a more functional approach of space use by different organisms belonging to the same food web.

The Diptera, Empididae

Diptera, Empididae are a good example of multi-habitat species. Diversity of species is high among this family in relation to their biological characteristics: feeding, mating habits or larval development (Burel *et al.*, 1998). For almost all the species edaphic larvae require nearly undisturbed soils to develop. The adults are predators or flower feeders. Empididae adults can be usually observed when they are alighted on or under shrub leaves. This phase can correspond to rest behavior or to predation behavior.

During the reproduction phase several species of Empidinae exhibit original and singular nuptial parades. Species of *Hilara* genus are more easily observed near or above water areas where they form mating swarms. Males hunt small prey by flying above water, they then embed their prey in silk and offer the balloons to females.

So Empididae individuals need different habitats in a given landscape to fulfil their life cycle: relatively undisturbed sites for larval development, shrub leaves for shelter sites, flower rich or insect rich sites for feeding and suitable swarming sites for breeding. These different habitats must be present within a given area in the landscape for the individuals to complete their life cycle in relation to their dispersal abilities (Morvan *et al.*, 1994).

Human activities can affect one or another of the stages of population survival by modification of habitat structure. We present some results on a key site in the *Hilara* life cycle, the breeding habitats which have been studied in the three hedgerow network landscapes described above.

During the adult activity period, *e.g.* from May to August, all types of open water areas (brooks, temporary ponds) have been surveyed in the three landscapes. All the swarms were located above brooks in various micro-climatic conditions (Morvan, 1996).

The brooks characterization has been realized according to the vegetation structure along verges. It considers the bushes at the brook and the presence of a wooded bank on one or both sides (Fig. 7). Bushy means that shrub vegetation such as brambles or willows covers the brook bed. Those parameters have been identified because they are known to influence swarm making by determining sunny spots on the water.

Figure 7. Example of the characterization of a stream structure using bushy and number of wooded banks.

For a second time the adjacent land-use has been considered to study the relation between water corridor structure and agricultural activities. On the three landscapes studied land use adjacent to brooks is mainly meadows. In site A woodlots and fallow land are also well represented, while in site C maize crops and other crops cover 24% of the area. Those differences may be related to the farming systems in these landscape sites. Opening of the hedgerow network is related to changes in farming activities and thus on changes in land cover (Table 3).

Table 3. Adjacent land use of streams in three hedgerow network landscapes. Site name is in relation to hedgerow density (site A: dense, site B: intermediate, site C: open)

	Site A	Site B	Site C
Permanent meadow	40.0%	60.4%	45.0%
Maize crop	12.8%	4.3%	13.2%
Other crops	2.6%	6.7%	10.6%
Wood & fallowland	37.1%	22.6%	20.7%
Other landuse	7..5%	6.0%	9.8%

Management of brook corridors varies in the different study sites. Figure 8 shows that the importance of bushy vegetation over the brooks decreases from the dense hedgerow network landscape (site A) to the more open one (site C). In site A the brook was for most of its length bordered by two wooded banks, Site C was heterogeneous, with brook sections that were covered after reallotment and drainage by farmers or without water during sampling period .

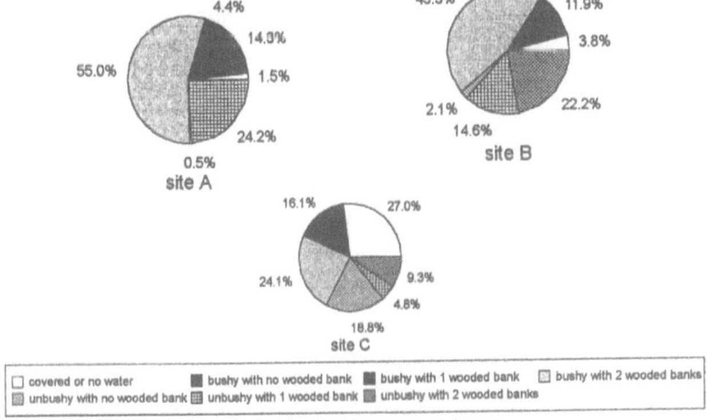

Figure 8. Stream structure in three landscapes.

The spatial distribution of swarms along the brooks showed a strong relationship between swarm location and brook vegetation structure. Indeed most of the swarms have been found over non-bushy brooks with one wooded bank whatever the site (Table 4).

Table 4. Swarm distribution at streams in three landscapes

	Site A	Site B	Site C	Total
Unbushy with no wooded bank	17	3	33	21
Unbushy with 1 wooded bank	75	77	50	67
Unbushy with 2 wooded bank	0	0	0	0
Unbushy with no wooded bank	8	0	0	0
Unbushy with 1 wooded bank	0	10	0	3
Unbushy with 2 wooded bank	0	10	17	9

This could be explained by the movement of insects searching for this favorable habitat. At a local scale, a swarm corresponds to coordinated flight of a great number of adults and this behavior could only be possible in open habitat. At a larger scale, the perception of swarms and the movement for each individual to reach the swarm could be difficult if there are two wooded banks which would act as barriers (Morvan, 1996).

The case of *Hilara* flies illustrates the fact that water corridors can be of great value for preservation of insects in agricultural landscapes. Their efficiency depends on the way they are managed and is related directly and indirectly to the adjacent land use. Directly, if proximity of suitable larval sites plays a role, indirectly as hedgerow structure and management are controlled by farming practices in adjacent fields.

The two Dipteran families considered above are quite different in their life history traits and the landscape features they are using at the larval or adult stage. However, common landscape features appear to influence their spatial distribution and their choice of particular places to express their behavior. In any case, areas in close proximity to brooks or other water bodies appear to be key sites which control dispersal processes and swarming through the nature of adjacent land mosaic and quality of hedges or river banks.

These zones are also considered important for the control of nutrient fluxes, as they may act as buffer zones if covered with permanent vegetation (Haycock *et al.*, 1997). To play a role in the maintenance of Dipteran communities, permanent grassland is far better than forested vegetation. We demonstrate, with this group, that the absence of a key element in a landscape can jeopardizes the whole life cycle of a species and may lead to its extinction.

2.3.2 STENOTOPIC SPECIES SPENDING THEIR ENTIRE LIFE CYCLE WITHIN ONE TYPE OF HABITAT

Some species disperse in the agricultural landscape without using the mosaic of fields. They are restricted to the network of uncultivated landscape elements during their whole life cycle. Some forest carabid beetles are able to live in agricultural landscapes at considerable distances from forests as long as a dense network of hedgerows remains (Burel, 1989). These species can survive in woody networks as fragmented populations. Local populations are located in small woods and large intersections of hedgerows. Linear features providing a minimal tree cover can be used as dispersal corridors linking populations (Petit & Burel, 1993). Extinction and recolonization occur according to landscape structure. The distribution of species at the landscape level is highly related to the continuity of the woody cover, the most isolated suitable habitats being less likely to be colonized (Petit, 1994). Figure 9 summarizes the spatial distribution of fragmented populations of *Abax parallelepipedus* in the hedgerow network.

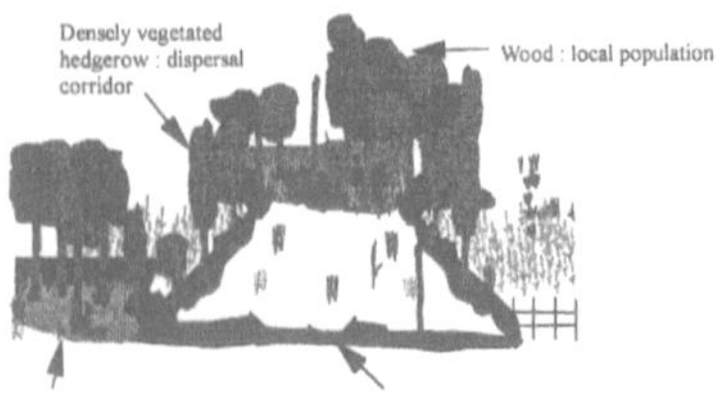

Figure 9. Representation of the fragmented populations of *Abax parallelepipedus* in a hedgerow network.

The structure of the woody network affects the dynamics of the species at the landscape level. A network with too many gaps will no longer be able to support forest carabids. But at a finer scale, the quality of linear features also clearly affects the survival of the individuals.

The corridor role of a woody linear element depends upon its size and structure. It will only exist if there is a dense herbaceous layer and shade provided by woody plant species. Hedgerow management is thus a key factor for the movements of carabids.

A long-term survey of individual movements was performed using radio-tracking techniques on the forest carabid, *Abax parallelepidedus* (Charrier *et al.*, 1997). The movements were compared in four types of landscape elements used by the species: a wood, a lane formed by two parallel hedgerows, a hedgerow with continuous tree cover, and a hedgerow with sparse trees. The elements also differed by the adjacent land-uses, either meadows or crops.

The walking pattern in the four elements was similar (random walk) but intensity of movements differed significantly (mean distance in 48 hours: ANOVA, $F = 4.306$, $p = 0.013$) (Fig. 10). Mean distances covered per 48 hours as well as the total area occupied by all individuals during the study were highest in the wood and decreased as vegetation cover decreased in linear features. Among the three linear features, lanes were the most efficient corridors (Table 5).

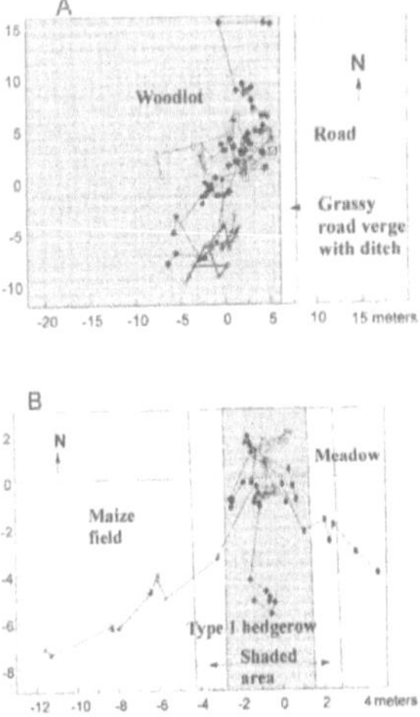

Figure 10. Movement trajectories for individuals traced in the woodlot ((a), n = 8 individuals) and in the densely vegetated hedgerow ((b), n = 7 individuals). (one spot = one individual).

Table 5. Mean values of distance covered in 48h and total area covered in two months by all the individuals traced in each of the four habitats (n= number of individuals, mean distance is given in m)

Habitat	n	Mean distance in 48h	Area covered (m²)
Wood	8	1.25±0.46	200
Lane	8	1.05±0.75	80
Hedgerow 1	6	0.45±0.16	20
Hedgerow 2	8	0.77±0.31	14

The nature of adjacent land-use, determining the sharpness of the transition between the uncultivated element and the agricultural matrix had an effect on the behavior of the beetle. Half of the carabids left linear features to enter meadows while only one beetle entered young maize crops with much bare soil. This suggests that the edge between the hedgerow and the meadow, which is a gradual environmental transition, is rather permeable for *A. parallelepipedus* while a sharp contrast in environmental conditions acts as a barrier.

In meadows, most beetles remained in the grassy strip along the hedgerows suggesting that this strip could be part of the functional corridor. As a consequence, agricultural practices in the meadows had a large effect on the survival of individuals. (Table 6). Mowing of one of the meadows during the study resulted in the loss of all beetles occurring in the adjacent grassy strip.

Table 6. Mortality in the different habitats. Mortality is 100 * (number of dead individuals found / total number of location in a given habitat) for the whole study period. n = number of individuals traced

Habitat	n	Mortality
Woodlot	8	0
Lane	15	2.4
Type 1 hedgerow	13	4.9
Type 2 hedgerow	17	5.8

The dynamics of forest carabid beetles in hedgerow networks is dependent on landscape structure as well as on the quality of woody landscape elements. Modifications of agricultural systems and practices can affect the survival of these species.

The quality of hedgerows is often altered as shown for the municipality of Saint Marcan (Fig. 11). Hedgerow length was not too strongly reduced between 1952 and today but tall and large hedgerows with a continuous tree cover (type 1 hedgerows) which represented 85% of the network in the fifties represent less than 40% today. They progressively turned into narrow hedgerows of lesser quality, exhibiting sparse trees (type 2 hedgerows).

For forest carabid beetles, which use hedgerows as corridors, such modifications of structure and quality of the woody network increase the functional isolation between local populations. For example the modifications of the woody network in Saint Marcan described above led to an increase of one third in the mean functional distance between two neighboring populations of *A. parallelepipedus*. Only some populations became totally isolated (no connections) but most populations became more isolated because lanes and densely vegetated hedgerows were replaced by corridors of lesser quality.

The case of forest ground beetles illustrates the fact that corridors can be of great value for the preservation of insects in agricultural landscapes. Special attention should be paid to maintain a connected network of woody habitats providing suitable conditions for these species. These linear features should not be managed too drastically and a continuous cover of shrubs and trees has to be ensured. In this way, extinctions of local populations can be partly prevented and recolonization processes can take place along the woody network.

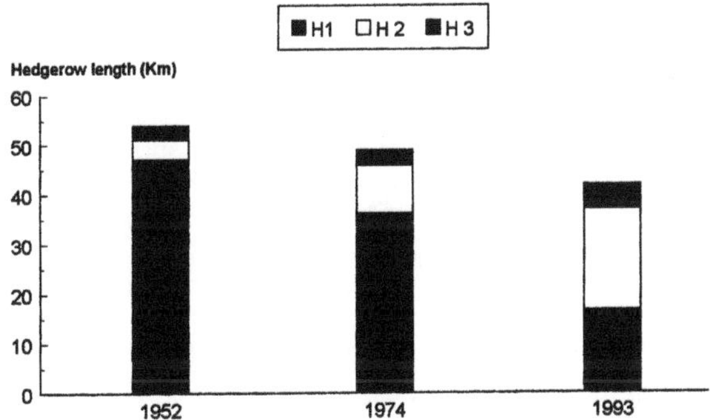

Figure 11. Changes in the hedgerow network of Saint Marcan - Brittany.

2.4 Modeling Insect Movements in Dynamic Landscapes

Use of simulation models is an important perspective for research at the landscape level. The scale of approach, the high variability and heterogeneity encountered makes it impossible to replicate experiences at this scale. Due to the wide range of technical, sociological, environmental factors driving landscape dynamics it is also very difficult to infer mechanisms from well-designed comparative studies (Baudry *et al.*, 1996). For obvious reasons, experimental tests are difficult to conduct at broad scales of human-defined landscape. The use of micro-landscapes for small scale experiments is advocated to study some specific mechanisms (Wiens & Milne, 1989). This may be of interest only for very local and well-defined processes, to scale up to the human landscape scale is highly speculative due to the high complexity of landscape organization. To extrapolate across scales would require knowledge about rules for translating, for example, individual space use to population dynamics, or from vegetation structure in a square meter to farm mosaic on a square kilometer. Our current lack of knowledge and theory prevents us from doing so.

Simulation models based on modeled dynamic landscapes, inferred from field studies and interviews, and on animal individual movements seem a way to cope with some of the difficulties encountered in landscape ecology. In our team we chose to develop a multi-agent simulation model for individual movement and dispersal in a changing landscape.

Simulation of dispersal in heterogeneous landscape is often based on an approach which models the movement of individuals (Wiens *et al.*, 1997; With *et al.*, 1997; Dunning *et al.*, 1995) in a spatially explicit environment. Most of the time, landscape structure does not vary during the simulation process. This is unrealistic in such dynamic landscapes as agricultural ones. Another question that may be addressed is the validity of simulations at a given scale while the mechanisms involved in species' movements operate at various scales.

The development of multi-agent simulation models seems promising to overcome these difficulties. With such tools the behavior of each agent is defined at its own operating scale; agents may be either landscape elements or organisms. For example pruning hedgerow trees occurs every twelve years, crop succession is annual and beetles may be modeled over two-day periods. Simulation of the dynamics of these different elements may be performed linked with GIS (Martin, *et al.*, 1997). These methods allow the complexity and the dynamics of agricultural landscapes to be taken into account.

2.4.1 AN EXAMPLE OF SIMULATION PROCEDURE: INDIVIDUAL MOVEMENTS OF *A. PARALLELEPIPEDUS* IN A HEDGEROW NETWORK LANDSCAPE

To model the individual behavior of *A. parallelepipedus* in a landscape we use a multi-agent system. It is defined as a set of agents interacting in a common environment. Each agent is an entity living in this environment, is autonomous and may modify the environment and itself. The system includes three main classes of carabid agents, *i.e.* sub-adults, adults and eggs, and agents that are landscape elements located on a digitized map.

Information on carabid agents includes:

1) mortality and reproduction: parameters are from literature data (Chaabane *et al.*, 1996)
2) individual movement within the landscape: movement depends on the habitat suitability and on a set of transition probabilities. The parameterization of this movement process is done according to the results of the radio-tracking study (Charrier *et al.*, 1997).
3) carrying capacity for landscape elements: as few data are available on this parameter at the landscape scale, we pose that as soon as the density of the animals exceeds the value of 0,5 individuals/m^2 (estimate density for the breeding sites (Chaabane *et al.*, 1996)), the probability of transition between two different habitats is set to one.

The carabid agents are located on a landscape map where hedgerows, woodlots, pastures, and crops are identified. Effects of landscape structure are assessed by running simulations on constructed maps derived from maps of our studied landscape units. To get a value of the network density lower than the current one, the initial network is changed by stochastically deleting the hedgerows using a uniform random generator.

This approach allows us to model the effect of some landscape parameters such as density of the hedgerow network or quality of the hedgerow vegetation structure on the survival of *A. parallelepipedus* at the landscape level.

Our objective for the near future is to integrate landscape dynamics within the model. As we have seen above, landscape elements are dynamic at their own spatio-temporal scale. The knowledge of farming practices, productions, and policies, permits prediction of dynamics. This is of overriding importance in rapidly changing landscapes where inferences from models that incorporate equilibrium assumptions may be highly misleading.

2.5 Discussion

Survival of insects in fine grain heterogeneous landscapes depends, for most of them, on movements between different landscape elements either of similar or different structure. The central hypothesis developed in landscape ecology is that the spatio-temporal dynamics of the landscape mosaic control insect population dynamics. In this paper we have presented some results that support this hypothesis.

In agricultural landscapes, farming systems are the key to explaining landscape organization at different hierarchical levels. Technical choices (machinery, crop versus grassland, ...) were responsible for the drastic changes in landscape structure in western France during the last decades. Introduction of new crops, particularly maize, within the farming system and crop succession explain the organization of the land use and land cover mosaics. Management techniques used by farmers on field boundaries are closely related to the adjacent crop or grassland and thus intimately linked to the farming system.

Over the world, farming systems are very different in both their size and spatial patterns of farms and the techniques employed by farmers. Our study sites are characterized by relatively small fields (1 to 10 ha) and farms (a few tens of ha), the fields of most farms being scattered over the landscape. Nevertheless, the general principles of landscape patterns being driven by farming activities should apply in most situations. Up to now, agronomists have not developed a theory to elucidate the mechanisms producing the spatial patterns. This is certainly an exciting field of research which, as it develops, will enable ecologists to incorporate the dynamics of farmed landscapes into the design of their empirical studies as well as into their models.

This type of research would also have practical implications for the design of agricultural-environmental policies as it would highlight the mechanisms to use to reach ecological targets, these mechanisms being at field, farm or set of farms level.

The different levels of organization we recognize in a landscape influence insect movement and spatial distribution. Quality of hedgerow vegetation, as managed by farmers, determines Chironomidae resting sites, influences *A. parallelepipedus* movement intensity and quality, and permits or inhibits swarm formation for *Hilara* species. Organization of the agricultural mosaic, and more specifically for the examples presented here, importance of grassland versus cropland plays a role in the dispersal phase of Empididae adults from larval sites to swarming sites. Connectivity of hedgerows at the landscape level is necessary for long time metapopulation survival of many carabid forest species, and hedgerow network density controls the dispersal fluxes of aquatic Chironomidae.

These examples confirm the interest of studying landscapes to manage insect populations in farmland. They also point out the multiplicity of responses to a given situation due to the biological characteristics of the different species. This could be considered as a major constraint for applying landscape ecology principles to conservation biology or pest management. For example a "good quality" hedgerow for a forest carabid beetle has a dense herbaceous layer and a continuous tree layer and its role as a dispersal corridor enhances the population viability. On the other hand it will act as a barrier for butterflies that need gaps in the canopy to move from one field to the other, if they are kept within one field small populations become isolated (Fry, 1995).

There is no unequivocal response of populations to landscape change. Burel *et al.* (1998) have shown that on a gradient of landscape openings in Brittany France, resulted, according to the different groups studied, either in a loss of species, a maintenance of species richness by species replacement or no change at all in species present. Invertebrate groups were all in the first or second class cited, therefore more sensitive to landscape change than vertebrates or plant species. To manage landscapes to enhance biodiversity or for integrated pest management necessitates understanding, more widely, the relationships between landscape. Species simulation studies based on real landscapes and on functional groups of species, based on their life history traits, are promising tools. They permit us to consider the complexity in time and space of the landscape and to assess species behavior given their biological traits.

Acknowledgments

We thank Dominique Vollant, Valérie Adamandidis and Manuel Simonot for their help during the field work, the identification of beetles and the sorting of Diptera. Daniel Denis provided help for the design and the management of the databases. This research has been financially supported by the Ministère de l'Environnement (Comité Ecologie et Gestion du Patrimoine Naturel), the European Commission (Programme Fair IV), the CNRS, Comité Systèmes ruraux du Programme Environnement.

2.6 References

Agger, P. & Brandt, J., 1988. Dynamics of small biotopes in Danish agricultural landscapes. *Landscape Ecology* **1**: 227-240.

Allen, T.H. & Hoekstra, T.W., 1992. *Toward a unified ecology*. Columbia University Press.

Altieri, M.A., 1983. Vegetational designs for insect habitat management. *Environmental Management* **7**: 3-7.

Baker, W.L., 1989. A review of models of landscape change. *Landscape Ecology* **2**: 111-135.

Bannister, N.R. & Watt, T.A., 1994. Hedgerow management: past and present. In: Watt, T.A. & Buckley, G.P. (Eds.), *Hedgerow management and nature conservation*. Wye College Press, pp. 7-15.

Barr, C.J., 1993. *Countryside survey 1990. Main report*. DOE.

Baudry, J., 1992. Dependance d'échelle d'espace et de temps dans la perception des changements d'utilisation des terres. In: Auger, P., Baudry, J. & Fournier, F. (Eds.). *Hiérarchies et échelles en écologie*. Naturalia Publications, pp. 101-113.

Baudry, J., Laurent, C. & Denis, D., 1997. The technical dimension of agriculture at a regional scale: methodological considerations. In: Laurent, C. & Bowler, I. (Eds.), *CAP and the regions: Building a multidisciplinary framework for the analysis of the EU agricultural space*. INRA Editions, pp. 161-173.

Baudry, J., Steyaert, P., Thenail, C., Deffontaines, J.P., Maigrot, J.L., Leouffre, M.C., Santucci, P. & Balent, G. 1996. *Approche spatiale des systèmes techniques agricoles et environnement*. Nouvelles fonctions de l'agriculture et de l'espace rural: enjeux et défis identifiés par la recherche, Toulouse, INRA Paris.

Bennett, A.F., 1990. Habitat corridors and the conservation of small mammals in a fragmented forest environment. *Landscape Ecology*, **4**: 109-122.

Boatman, N., (Ed.), 1994. *Field margins: integrating agriculture and conservation*. British Crop Protection Council.

Bunce, R.G.H. & Hallam, C.J., 1993. The ecological significance of linear features in agricultural landscapes in Britain. In: Bunce, R.G.H., Ryszkowski, L. & Paoletti, M.G. (Eds.), *Landscape ecology and agroecosystems*. Lewis. pp. 11-20.

Burel, F., 1989. Landscape structure effects on carabid beetles spatial patterns in Western France. *Landscape Ecology* **2**: 215-226.

Burel, F. & Baudry, J., 1990. Structural dynamic of a hedgerow network landscape in Brittany France. *Landscape Ecology* **4**: 197-210.

Burel, F. & Baudry, J., 1995a. Understanding biodiversity in rural areas: species spatial distribution in changing landscapes. *Agriculture Ecosystems and Environment* **55**: 193-200.

Burel, F. & Baudry, J. 1995b. Social, esthetical and ecological aspects of hedgerow in rural landscapes as a framework for greenways. *Landscape and Urban Planning* **33**: 327-340.

Burel, F., Baudry, J., Butet, A., Clergeau, P., Delettre, Y., Le Cœur, D., Dubs, F., Morvan, N., Paillat, G., Petit, S., Thenail, C., Brunel, E. & Lefeuvre, J.C., 1998. Comparative biodiversity along a gradient of agricultural landscapes. *Acta Oecologica* **19**: 47-60.

Chaabane, K., Loreau, M. & Josens, G., 1996. Individuals and population energy budgets of *Abax ater* (Coleoptera, Carabidae). *Animal Zoology Fennici* **33**: 97-108.

Charrier, S., Petit, S. & Burel, F., 1997. Movements of *Abax parallelepipedus* (Coleoptera, Carabidae) in woody habitats of a hedgerow network landscape: a radio-tracing study. *Agriculture, Ecosystems & Environment* **61**: 133-144.

Crumley, C.L. & Marquardt, W.H., (Eds.), 1987. *Regional dynamics. Burgundian landscapes in historical perspective*. Academic Press, Inc.

Deffontaines, J.P., Thenail, C. & Baudry, J., 1995. Agricultural systems and landscape patterns: how can we build a relationship? *Landscape and Urban planning* **31**: 3-10.

Delettre, Y.R., 1984. *Recherches sur les Chironomides (Diptera) à larves édaphiques: biologie, écologie, mécanismes adaptatifs*. Université de Rennes **1**: 310.

Delettre, Y.R. & Lagerlöf, J., 1992. Abundance and life-history of terrestrial Chironomidae (Diptera) in four Swedish agricultural cropping systems. *Pedobiologia* **36**: 69-78.

Dover, J.W., 1991. *The conservation of insects on arable farmlands. The conservation of insects and their habitats. 15th Symposium of the Royal Entomological Society of London*, pp. 293-318.

Dowdeswell, W.H., 1987. *Hedgerows and Verges*. Allen & Unwin Ltd.

Duelli, P., M., Studer, M., Marchland, I. & Jakob, S., 1990. Population movements of arthropods between natural and cultivated areas. *Biological Conservation* **54**: 193-207.

Dunning, J.B., Stewart, D.J., Danielson, B.J., Noon, B.R., Root, T.L., Lamberson, R.H. & Stevens, E.E., 1995. Spatially explicit population models: current forms and future uses. *Ecological Applications* **5**: 3-11.

Ehanno, B., 1988. *Les hétéroptères mirides de France (Heteroptera:Miridae): distribution biogéographique et contribution à l'étude de leurs rapports avec les plantes et les milieux naturels.* Thesis d'état. Rennes -France.

Fahrig, L. & Merriam, H.G., 1985. Habitat patch connectivity and population survival. *Ecology* **66**: 1762-1768.

Fittkau, E.J., Reiss, F. & Hoffrichter, O., 1976. A bibliography of the Chironomidae. *Gunneria* **26**: 1-177.

Fry, G., 1995. Landscape Ecology and insect movement in arable ecosystems. In: Glen, D.M., Greaves, M.P. & Anderson, H.M. (Eds.), *Ecology and integrated farming systems*. John Wiley & Sons. pp. 177-202.

Glen, D.M., Greaves, M.P. & Anderson, H.M., (Eds.), 1995. *Ecology and integrated farming systems.* John Wiley & Sons.

Gustafson, E.J. & Gardner, R.H., 1996. The effect of landscape heterogeneity on the probability of patch colonization. *Ecology* **77**: 94-107.

Haycock, N.E., Burt, T., Goulding, K.W.T. & Pinay, G. 1997. *Buffer zones: their processes and potential in water protection.* Quest Environment Publisher.

Hegarthy, C.A., McAdam, J.H. & Cooper, A., 1994. Factors influencing the plant species composition of hedges - implications for management in Environmentally Sensitive Areas. In: Boatman, N. (Ed.), *Field margins integrating agriculture and conservation.* British Crop Protection Council: 227-234.

Hoffrichter, O., 1981. Supplement 1 to "A bibliography of the Chironomidae". *Gunneria* **37**: 1-8.

Hudson, P.L., 1987. Unusual larval habitats and life history of Chironomid (Diptera) genera. *Entomologica Scandinavica* **29**: 369-373.

INRA, ENSA & Rennes, U.d., 1976. Les bocages: Histoire, Ecologie, Economie. Table ronde CNRS: les écosystèmes bocagers, Rennes, INRA-Rennes.

Laurent, C. & Bowler, I., (Eds.), 1997. *CAP and the regions: Building a multidisciplinary framework for the analysis of the EU agricultural space.* INRA Editions.

Laurent, C., Langlet, A., Chevallier, C., Jullian, P., Maigrot, J.L. & Ponchelet, D., 1994. Ménages, activités agricoles et uilisation du territoire: du local au global à travers les RGA. Cahiers d'études et de recherches francophones. *Agricultures* **3**: 93-107.

Le Cœur, D., Baudry, J. & Burel, F., 1997. Field margin plant assemblages: variation partitioning between local and landscape factors. *Landscape and Urban Planning* **37**: 57-72.

Legendre, L. & Legendre, P., 1984. L'interprétation des structures. *Ecologie numérique. 2 La structure des données écologiques.* Masson, Presses de l'Université de Quebec. **2**: 160-204.

Luginbühl, Y., 1995. Des formes d'arbres insolites. *Pour la science*: 8-9.

MacArthur, R.H. & Wilson, E.O., 1967. *The theory of island biogeography.* Princeton University Press.

Martin, M., Burel, F., Petit, S. & Rodriguez, R., 1997. Invertebrate populations in a hedgerow network landscape: an individual based model- conceptual and methodological approach. In: Cooper A. & Power J. (Eds.), *Species dispersal and land use processes.* Coleraine, Northern Ireland, UK IALE.

Meeus, J.H.A., 1995. Pan-European landscapes. *Landscape and Urban Planning* **31**: 57-79.

Merriam, G., 1988. Modeling woodland species adapting to an agricultural landscape In: Schreiber K.F. (Ed.), Connectivity in Landscape Ecology. *Münstersche Geographische Arbeiten*. pp. 67-68.

Merriam, H.G., 1984. Connectivity: a fundamental characteristic of landscape pattern. In: Brandt, J. & Agger, P. (Eds.), *Methodology in landscape ecological research and planning.* Roskilde University Centre, Denmark. **1**: 5-15.

Meyer, J., 1972. L'évolution des idées sur le bocage en Bretagne. In: La pensée géographique française contemporaine. Presses universitaires de Bretagne: 453-467.

Meyer, W.B. & Turner II, B.L. (Eds.), 1994. *Changes in land use and land cover: a global perspective.* Cambridge University Press.

Morant, P., Le Henaff, F. & Marchand, J.P., 1995. Les mutations d'un paysage bocager: essai de cartographie dynamique. *Mappemonde* **1**: 5-8.

Morgan Evans, D., 1992. Hedges as historic artifacts. In: Watt, T.A. & Buckley, G.P. (Eds.), *Hedgerow management and nature conservation.* Wye College Press. pp. 107-118.

Morvan, N., 1996. *Structure et Biodiversité de paysages de bocage: Le cas des empidides (Diptera, Empidoidea).* Thesis. U.F.R. Sciences de la Vie et de l'Environnement. Université de Rennes 1, Rennes. Rennes **1**: 215.

Morvan, N., Delettre, Y.R., Trehen, P., Burel, F. & Baudry, J., 1994. *The distribution of Empididae (Diptera) in hedgerow network landscapes.* Field margins integrating agriculture and conservation. *BCPC monograph* **58**: 123-127.

Odum, E.P. & Turner, M.G., 1990. The Georgia landscape: a changing resource. In: Zonneveld, I.S. & Forman, R.T.T. (Eds.), *Changing landscapes an ecological perspective* **28**.

Petit, S., 1994. *Metapopulations dans des réseaux bocagers: analyse spatiale et diffusion.* Thesis. Université de Rennes I: 138.

Petit, S. & Burel F., 1993. Movement of *Abax ater* (Col. Carabidae): do forest species survive in hedgerow networks? *Vie et Milieu* **43**: 119-124.

Pollard, E., Hooper, M.D. & Moore, N.W., 1974. *Hedges.* W. Collins and Sons.

Rackham, O., 1986. *The history of the countryside.* J.M. Dent & Sons Ltd.

Schippers, P., Verboom, J., Knaapen, J.P. & van Apeldoorn, R.C., 1996. Dispersal and habitat connectivity in complex heterogeneous landscapes: an analysis with GIS-based random walk model. *Ecography* **19**: 97-106.

Seddon, A.M., 1986. Abundance and life-history of four species of terrestrial Chironomidae (Diptera) from deciduous woodland soil in south-east England. *Entomologist Monthly Magazine* **122**: 219-228.

Sendstad, E., Solem, J.O. & Aagaard, K., 1977. Studies of terrestrial chironomids from Spitsbergen. *Norwegian Journal of Entomology* **24**: 91-98.

Solbrig, O.T., 1991. *From genes to ecosystems: a research agenda for biodiversity.* IUBS-SCOPE-UNESCO.

Stamps, J.A., Buechner, M. & Krishnan, V.V., 1987. The effect of edge permeability and habitat geometry on emigration from patches of habitat. *American Naturalist* **129**: 533-552.

Strenzke, K., 1950. Systematik, morphologie und okologie der terrestrischen Chironomidae. *Archiv für Hydrobiologie Supplement* **18**: 207-414.

Thenail, C.B.J., 1996. *Consequences on landscape pattern of within farm mechanisms of land use changes (example in western France). Land use changes in Europe and its ecological consequences.* Tilburg, European Center for Nature Conservation.

Thomas, M.B., Wratten, S.D. & Sotherton, N.W., 1992. Creation of "island" habitats in farmland to manipulate populations of beneficial arthropods: predator densities and species composition. *Journal of Applied Ecology* **29**: 524-531.

Verboom, B. & van Apeldoorn, R., 1990. Effects of habitat fragmentation on red squirrel, *Sciurus vulgaris* L. *Landscape Ecology* **4**: 171-176.

Vermeulen, J.W., 1995. *Road-side verges: Habitat and corridor for carabid beetles of poor sandy and open areas.* Wageningen.

Watt, T.A. & Buckley, G.P., (Eds.), 1994. *Hedgerow management and nature conservation.* Wye College Press.

Wiens, J.A. & Milne, B.T., 1989. Scaling of landscapes in landscape ecology, or landscape ecology from a beetle's perspective. *Landscape Ecology* **3**: 87-96.

Wiens, J.A., Schooley, R.L. & Weeks, R.D.J., 1997. Patchy landscapes and animal movements: Do beetles percolate? *Oikos* **78**: 257-264.

Wiens, J.A., Stenseth, N.C., Van Horne, B. & Ims, R.A., 1993. Ecological mechanisms and landscape ecology. *Oikos* **66**: 369-380.

With, K.A., Gardner, R.H. & Turner, M.G., 1997. Landscape connectivity and population distribution in heterogeneous environments. *Oikos* **78**: 151-169.

Wratten, S.D. & Van Emden, H.F., 1995. Habitat management for enhanced activity of natural enemies of insect pests. In: Glen, D.M., Greaves, M.P. & Anderson, H.M. (Eds.), *Ecology and integrated farming systems*. John Wiley & Sons Ltd, pp. 117-145.

CHAPTER 3

THE USE OF INVERTEBRATES IN EVALUATING RURAL SUSTAINABILITY

MAURIZIO G. PAOLETTI
Dipartimento di Biologia Università di Padova, Padova, Italy

CARLOS MARTIN CANTARINO
Dept. Ecologia, Universidad de Alicante, Alicante, Spain

3.1 Introduction

Producing food in a more sustainable way is the target of most developed and developing countries. Several policies in the last few years have been adopted and could be used to promote sustainability such as:

1) Pesticide reduction (The Netherlands, Denmark, Ontario province in Canada, Sweden have made laws addressed at this target (Pimentel, 1997)).
2) Set aside of farmland in the European Community, United States, etc., to reduce surplus and/or compensate farmer incomes.
3) Promotion of some rules to protect biodiversity in delicate habitats near water captions, water basins and reservoirs, in the protection belt around national and regional parks, etc.
4) Reduction of urban and industrial impact such as reduction of fossil energy based activities, abatement of pollution, recycling, use of alternative low input practices, etc.

In most cases these policies have been promoted as potentially aimed at improving the environment and in most developing and developed countries the rural and natural landscapes are sometimes close to industrial areas or urbanized areas in a complex web.

Can biodiversity in general and of invertebrate species in particular, the most abundant living biota in the planet, be used to monitor change leading to higher sustainability in rural landscapes? In the last few years several theoretical and field studies seem to provide evidence that this may be true.

The landscape is changing and patterns are related both to the recent and the remote past. Some intriguing elements such as large animals (cows, sheep, goats, and pigs) and cereals (wheat, oats, rye, and barley) have their origins from a restricted area, especially the Fertile Crescent, which nowadays is almost desert (Paoletti, 1997). Our lifestyle and rural landscapes are based on these keystones with the addition of few others key plants (potatoes, corn, rice, and sugar beets). All these key species have transformed the structure and shape of our surroundings.

Human population, even if extremely high, is dependent on agricultural production for everyday food and welfare. However, fossil energy is a limited resource in both industrial and developing countries.

B. Ekbom, M. Irwin and Y. Robert (eds.), Interchanges of Insects, 33-52

Solar energy and the natural environment lend major support to agriculture. But in most countries, especially in the industrial ones, fossil energy based technology, (for example: different pesticides and fertilizers, irrigation, etc.), is responsible for the current high yields. For instance, about 17% of the energy consumed in the USA is linked to agriculture (Pimentel & Pimentel, 1996).

In this chapter we will focus on the invertebrates as a consistent part of biodiversity and as the tool for assessing rural landscapes and improving its sustainability.

To address this possibility we first discuss two points about landscape and sustainability.

3.2 What is Rural Landscape? Large Animals versus Small Animals?

Species assemblages, patterns, and composition have been consistently modified by the transformation from natural to rural landscapes in the past 8,000-12,000 years. For instance, large species of earthworms and ground beetles have been eliminated almost completely from most rural areas (Paoletti, 1998). Large animals, such as ruminants have replaced small vertebrates in range lands, which are now made up of a limited number of grasses. Perennials have been substituted by annuals, basic crops are usually annuals. Forests have been reduced (with their potential perennial trees like oaks and chestnuts in temperate countries or Sago palms and other starchy fruit palms in the tropics). In most transitions from hunter-gatherers to agricultural societies there is a loss of trees bearing fruit etc. to annual, short cycle plants.

Larger grains overcome small grains. For instance, it has been observed that traditional small grains like millet and sorghum in Africa or quinoa and amaranth in the Andean regions tend to be substituted by the larger cereals such as corn, wheat, or barley (NRC, 1989; BSTID, 1996; Paoletti, 1995). The same trend can be observed in the Mediterranean regions, where small grains such as millet, sorghum, panicum, small legumes such as *Lathyrus* sp. and vetch or small fruits such as *Crataegus azerollus*, *Sorbus domestica*, *Mespilus germanica*, etc. tend to be abandoned in favor of larger ones.

In addition, in the original areas of plant domestication; such as for potatoes the Andes; the original, native varieties have been rapidly replaced by the current western varieties. In the Andes of Venezuela, near Merida, for instance, seed potatoes mostly come from The Netherlands. In Ecuador pasture grasses come from Africa. Leguminous pastures in the Amazonas,Venezuela, tend to come from Africa as well. Most original crops are displaced by imported seeds and associated technologies.

Landscapes in industrial countries are a mixture of history, climate, natural ecosystems in a coevolutionary framework. Also the "primary" forest is sometimes a mixture of combinations of plants that have been affected by human intervention. For instance, in the Amazonas the Kayapo Indians or Piaroa actively disseminate seeds of useful "wild" plants trees in the forest (Posey, 1992 and M.G.P. personal observation, July, 1997). This activity is also a form of domestication. Cocona or tupiro (*Solanum sessiliflorum*) and similar edible plants are voluntarily disseminated in slash and burn cultivation and in household gardens through human defecation (J. Salick, personal communication and M.G.P. personal observation). In the European hedgerows seeds are disseminated by bird defecation.

These many processes are involved with animal and crop selection and association. For instance, the dimension of the fields has some links with the animal draft power, the plant associations in pastures with their ruminant animals. The domesticated large animals have consistently shaped the landscapes in most areas around us. For instance trampling, use of fire, erosion, and reduction of woodlands are in most cases linked with this historical option.

3.3 What is Sustainability?

Empirically speaking sustainability is a local concept depending on the mix of environment, economics and peoples in each region. Rather than a definitive set of options sustainability is a flexible target (Conwey & Barber, 1990) (Table 1). Over time this theoretical sustainable system should not be degraded but this is difficult to measure, especially in current agroecosystems. In any case, most agricultural based landscapes, lose soil (Pimentel *et al.* 1995) and species (Paoletti, 1998), and as assumed since the work of Carter and Dale (1974) most civilizations in the past have collapsed due to the poor use of soil and renewable resources. Domestication processes that began the shift to agriculture lead to an incredible transformation of the landscape. Increasing the dimension of the organisms to be cultivated, improving the efficiency of the large, previously wild animals. In this process most mini-livestock (especially terrestrial invertebrates, rodents, amphibians, and reptiles) and semi-domesticated plants disappeared as resources for humans. In the mountains of Friuli Venezia Giulia recollection of wild plants for preparing the dish Pistic was the rule, in the spring up to 52 species were collectively recollected. The more productive cereals made these plants become if not an abandoned at least a limited local resource (Paoletti *et al.* 1995). In the tropics most of the foods come traditionally from perennials and small animals (mini-livestock), so the strongest devastating effect on the forests in tropical Amazonas areas is the adoption of plants and animals coming from the Fertile Crescent (like cows, sheep, or goats). We could expect to promote the domestication process of the small animals once again. But the available knowledge is limited and traditions in food patterns are difficult to change in a short time span, especially for western peoples that progressively have eliminated the small creatures from their diets (Paoletti & Bukkens, 1997). Insects and small animals such as mice were eaten by Greeks and Romans, and even considered a delicacy in some cases (Beavis, 1988), but through Western history this kind of food has been socially banned and relegated to rural or somehow marginal social classes until its almost practical disappearance nowadays. However, in some countries like China this food is still important and there is local consumption of small invertebrates (crustaceans, mollusks, insects) and plants associated with ponds and rice or with silkworm production (Paoletti & Bukkens, 1997; Paoletti 1999a).

Table 1. Comparison of social, economic, and environmental sustainability (from different sources, but especially from Goodland & Pimentel, 1998)

Social Sustainability	Economic Sustainability	Environmental Sustainability
Cohesion of community, cultural identity, diversity, solidarity, tolerance, humility, compassion, patience, forbearance, fellowship, cooperation, fraternity, love, pluralism, commonly accepted standard of honesty, laws, discipline, etc. constitute the part of social capital least subject to rigorous measurement, but essential for social sustainability.	Economic capital should be stable. The widely accepted definition of economic sustainability is ***maintenance of capital***, or keeping capital intact. The amount consumed in a period must maintain the capital intact because only the interest rather than capital should be consumed.	Although ES is needed by humans and originated because of social concerns, ES itself seeks to improve human welfare by protecting the sources of raw materials used for human needs, and ensuring that the sinks for human wastes are not exceeded, in order to prevent harm to humans. Humanity must learn to live within the limitations of the biophysical environment. ES means natural capital must be maintained, both as provider of inputs of sources and as sink for wastes. This means holding the scale of the human economic subsystem to within the biophysical limits of the overall ecosystem on which it depends. ES needs sustainable consumption by a stable population.
This ***moral capital*** requires maintenance and replenishment by shared values and equal rights, and by community, religious and cultural interactions. Without such care it depreciates as surely as would physical capital.	Economics have rarely been concerned with natural capital (*e.g.* intact forests, healthy air, stable soil fertility). To the traditional economic criteria of allocation and efficiency must now be added a third, that of scale. The scale criterion would constrain throughput growth - the flow of material and energy (natural capital) from environmental sources to sinks.	On the sink side, this translates into holding waste emissions within the assimilative capacity of the environment without impairing it. On the source side, harvest rates of renewables must be kept within regeneration rates.
Human and social capital, investment in education, health and nutrition of individuals is now accepted as part of economic development, but the creation and maintenance of social capital as needed for social sustainability is not yet adequately recognized.	Economics values thing in money terms and valuing the natural intergenerational capital like soil, water, air, biodiversity is problematic.	

Rural landscapes are not an homogeneous assemblage of plants animals and soils but comprise a network of "natural" or modified structures such as field margins, hedgerows, or river and channel banks, woodlots, shelterbelts, woodlands, ponds, marshes, swamps, abandoned fields, gardens, etc. It is not clear but seems a constant that new colonizers, such as the Ancient Romans produced consistent landscape modifications by resettling forested areas, rearing trees, and introducing hedgerows in their rural modified landscape (Fig. 1, Riese Pio X). This was made in Italy but possibly also in other European areas such as in England (Fig. 2). Similar settling of the previously forested landscape was done in the United States and Canada in the last century. This operation and its evolution through the decreased amount of margins create consistent reduction of the small scale fragmentation and decreased presence of the mosaic effect.

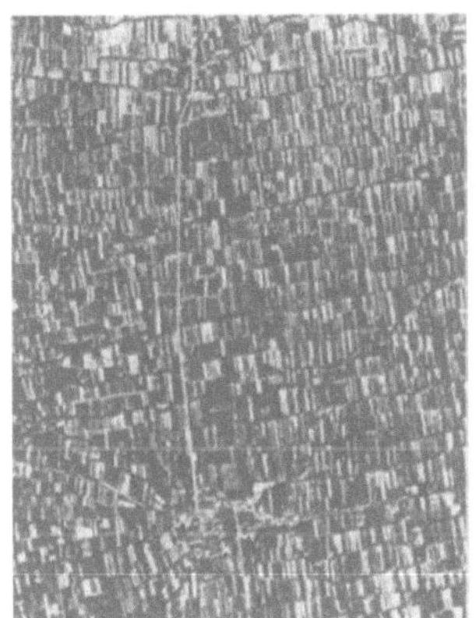

Figure 1. Centuriated area in northeastern Italy (Riese, Pio X, Treviso) showing the still persistent fingerprint of the ancient Roman centuriation on the territory. As in many countries in the last 60 years most landscapes have been severely transformed making larger fields from small ones, reducing trees and hedgerows and decreasing the margin effects.

Looking at the field 'per se' is not the way to measure biodiversity in the rural landscapes. We have to consider the mosaic in different parts. Movement, colonization, and recolonization among the different parts of the landscape are the rule rather than the exception. In many cases different, less disturbed areas, serve as recolonization sources for the more disturbed portions. Hedgerows, shelter belts, undisturbed margins, trees, woodlots, wild vegetation alongside lanes, roads, channels, and ditches can consistently provide sources for field colonization. Perennial crops such as alfalfa, hay, and orchards (low input) especially when covered by living mulches can be important sources for recolonization.

Figure 2. Centuriated areas in England.

3.4 Methodologies as a Crucial Point for Developing Expert Systems

To assess sustainability consistent tools have to be developed in order to assess change and compare different options in the landscape. One interesting tool is the use of invertebrates. Soil invertebrates have been suggested as key elements for soil formation and plant growth health (Dindal, 1989; Brussaard *et al.*, 1997). Soil biota have been suggested as tools in assessing different impacts and management strategies (Van Straalen, 1997; Paoletti & Bressan, 1996; Paoletti, 1998). Among invertebrates, many other taxa and animal guilds have been suggested as potential tools to assess, above ground, the landscape structure and function, for instance insects (Heliövaara & Väisänen, 1993; Paoletti, 1998, 1999b).

To assess the landscape insects and other terrestrial invertebrates can give an important indication being numerous, relatively known (but not very popular) and present in most situations. To make progress in using invertebrates as potential bioindicators (Paoletti, 1998) a consistent change of tools is required from the manuals to the computer managed expert systems. We expect that computer technology will make large inroads towards a better use of this invertebrate based tool to assess the landscapes around us.

3.4.1 APPLE ORCHARD ASSESSMENT (NORTHERN ITALY, NEAR BOLZANO)

Rural areas with intensive apple and grape cultivations are associated with some alpine valleys such as Val Tellina, Val di Non, Val Venosta, Val d'Adige in Italy. Intensive conventional apple orchards and vineyards need a quite high input of chemical fertilizers and pesticides, including high amounts of fungicides to control pathogenic fungi (Paoletti, 1997a). Most apple intensive areas have consistent problems with high pesticide input and alternatives such as resistant plants, efficient marketing of new low input varieties, are only at the beginning.

We have assessed, using bioindicators, one organic apple orchard (adopting shallow tillage in the surface and no fertilizer and pesticide input) comparing it with a nearby conventional apple orchard (high input, herbicides, pesticides and chemical fertilizers used) and, as a non-managed reference, one piece of deciduous woodland (Schweigl, 1989; Paoletti *et al.*, 1995). The bioindicator tools used were hand sorted earthworms and macro-invertebrates (operating hand sorting cores 30 x 30 cm, 30 depth with a spade). In addition pitfall traps were adopted. Five site repetitions were performed and almost monthly sampling done.

Both earthworms and some other groups collected by hand sorting gave an important indication of the different input systems (Fig. 3). In particular numbers of the large animal guilds were consistently higher in the organic orchard. In addition sampling with pitfall traps showed that at least carabids have been greatly reduced in the conventional apple orchard (Fig. 4).

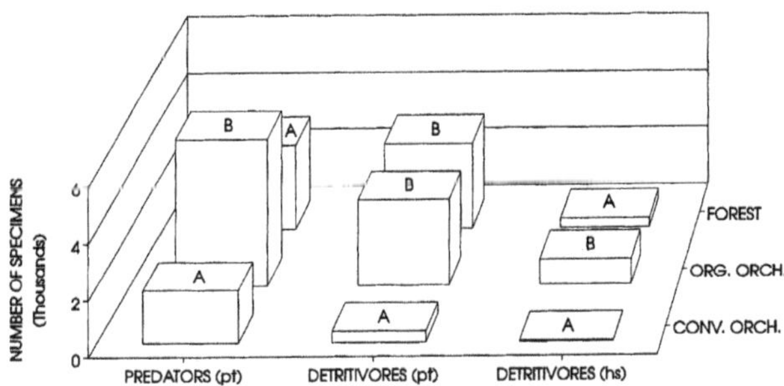

Figure 3. The detritivores as a composite group can sometimes, assessed as numbers, be sufficiently efficient in assessing different farming systems. Observe that the predators are not so affected. In this case one organic orchard versus one conventional orchard and one reference coppiced deciduous forest (From Paoletti *et al.*, 1995).

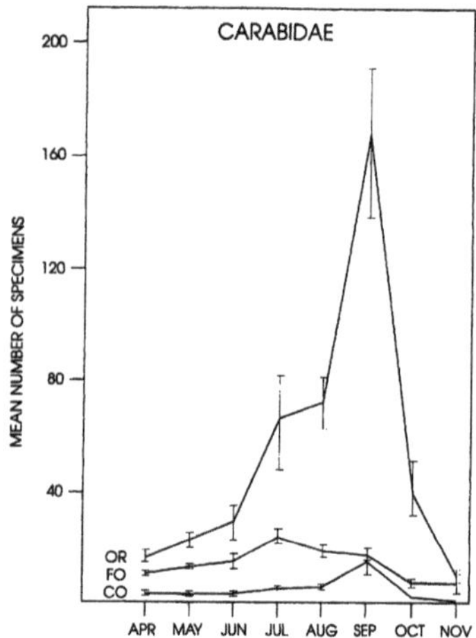

Figure 4. Carabids collected by pitfall traps are severely damaged by the conventional orchard high input farming (from Paoletti *et al.*, 1995).

In addition to the current input of pesticides (up to 72 kg/ha) in the conventional apple orchard residues of old pesticides have been accumulated, such as DDT metabolites and also arsenic, up to 50 ppm (Schweigl, 1989). Concentrations of these organochloride compounds can be degraded (such as ppDDT, opDDT, and ppDDT) only if a high microbial activity is found in the soil and there is adoption of organic mulching and appropriated incorporation of organic matter into the topsoil. These conditions are present in the organic orchard studied.

Which is the alternative to high input apple orchards? The key problems are pathogenic fungi especially scab and powdery mildew requiring high doses of fungicides. In addition codling moth and a few other Lepidoptera affect these crops requiring insecticides. To reduce this trend scab resistant varieties could be adopted and integrated or organic farming adopted. However to cope with these options clear policies and premiums for supporting organic productions must be implemented and better marketing strategies must be developed.

3.4.2 PEACH ORCHARDS (CENTRAL ITALY, NEAR FORLÌ)

We assessed, for two years, with the invertebrates, different peach orchards to evaluate the different farming systems (Paoletti *et al.*, 1993). We selected six farms: two organic, two integrated and two conventional peach orchards in one intensive orchard area in the Romagna region in Italy. The organic orchards had minimum pesticide input. In integrated farms only low toxicity chemical pesticides where used under scouting and threshold targets. Different sampling strategies have been adopted, from sweeping to pitfall traps to the entomological umbrella.

Figure 5 shows the species abundance reaction to the different farming strategies, in addition some taxa better than others respond to different farming practices. For instance the isopods, earwigs, and some carabid species singularly respond to the progressively higher farming impact (Fig. 6). Other species such as the spider *Oedothorax apicatus* react positively to the high input farming in the opposite way when compared to *Pachygnatha degeeri* (Fig. 7). In this case we could say that these species could be the interesting key bioindicators. However, selecting just a few key bioindicator species is not the best strategy to compare different environmental impacts because the few key species can disappear or be present just for non expected reasons. It would then be more useful to consider and assess a larger number of species inside one major taxonomic group or of different taxa.

Alternatives to reduce conventional inputs are: living mulches, resistant plants, marketing strategies to promote low input products, appropriate premium policies for low input farming, educational programs for consumers to appreciate products coming from low input farming.

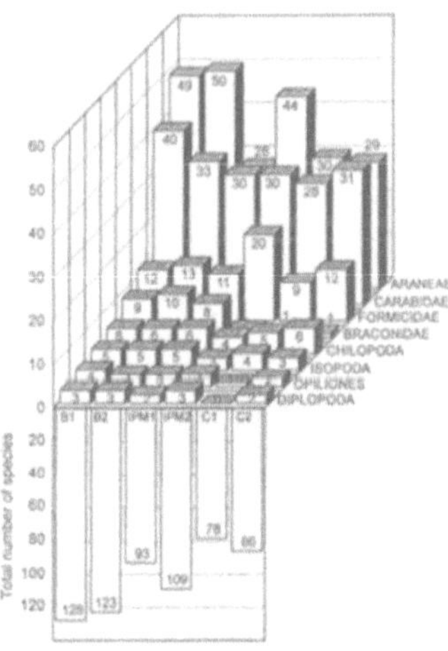

Figure 5. Peach orchards and invertebrate species abundance along with different farming systems. Number of arthropod species and input strategies in three peach orchards in Emilia Romagna, Italy. B1 and B2 are biological orchards; IPM1 and IPM2 are integrated orchards; C1 and C2 are conventional high input orchards. A decreased number of invertebrate species was noted in integrated and conventional farms compared to biological (organic) farms (From Paoletti & Sommaggio, 1996). Sampling was performed by pitfall traps and sweep nets on a monthly basis for two years.

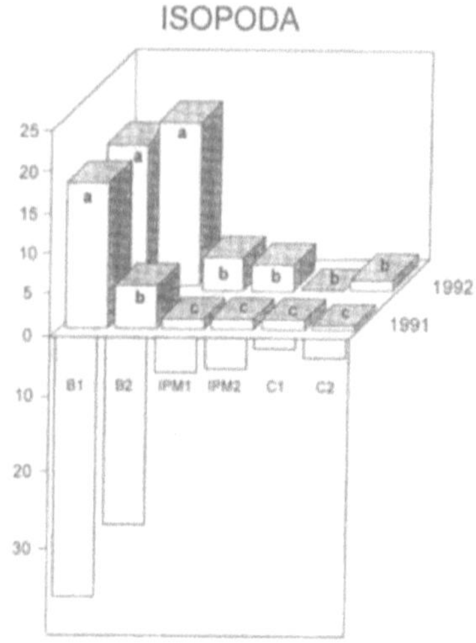

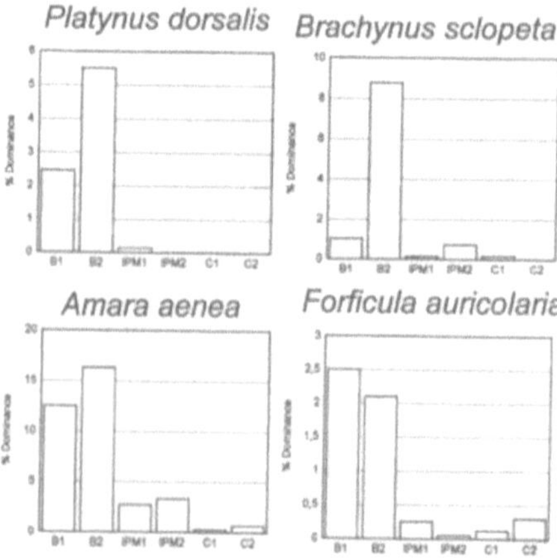

Figure 6. Isopods, Carabids and earwigs as indicators of different farming system impact as in the peach orchard situation (as in Figure 5).

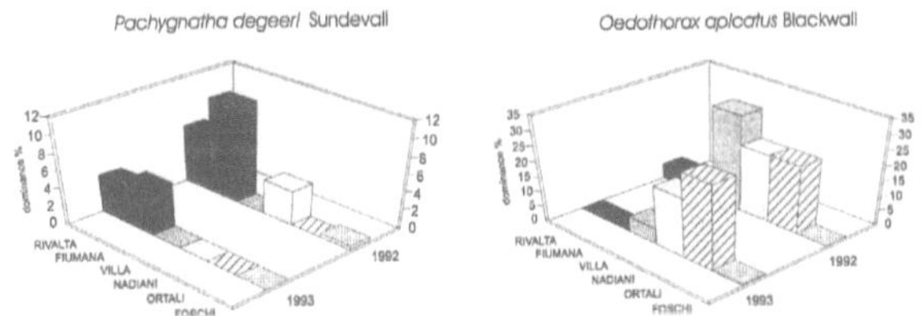

Figure 7. Two key spider bioindicators of impact in the peach orchard environment.

3.4.3 COMPARING DIFFERENT ORCHARDS (CENTRAL ITALY)

By reducing the number of taxa investigated as potential bioindicators to the earthworms alone we wanted to see if it was possible to minimize both the time consuming sampling and identification. From the low-plain in the north-east of Italy (Emilia Romagna) we selected and analyzed 64 ecosystems including vineyards and three types of orchards: apple, peach, and kiwi, characterized by different chemical inputs (Paoletti *et al.* 1998). We desired to assess comparatively these different orchards by using hand sorting of earthworms (cores 30 x 30 cm). There was a significant effect of both crop type and tillage on the biomass and abundance of total earthworms (Fig. 8). Cultivation operations in between the orchards rows reduced earthworm mean biomass by 42% in peach orchards, 36% in apple orchards, 20% in kiwi orchards, and 34% in vineyards. Earthworm mean abundance was reduced by 47%, 37%, 21%, and 64%, respectively. We found a significant, negative regression with copper and zinc content in the soil (Figs. 9, 10); the total earthworm variability explained by copper (expressed as r^2 in the regression analysis) was 50%.

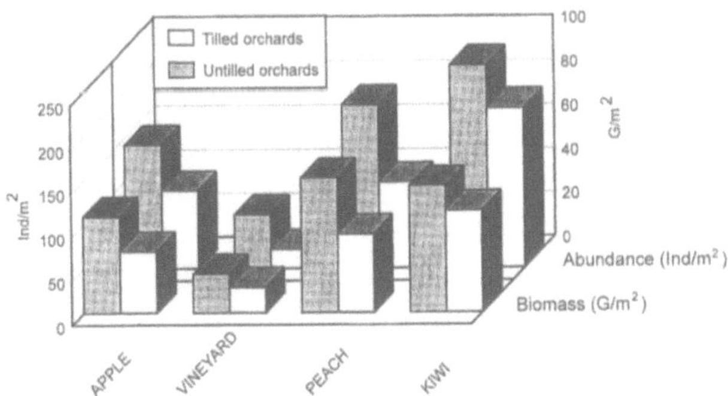

Figure 8. Total earthworms biomass and abundance in different orchards; tillage differences: $p < 0.0001$ for biomass and abundance; tillage treatment effect: $p < 0.0001$ for biomass and abundance (Kruskall-Wallis test).

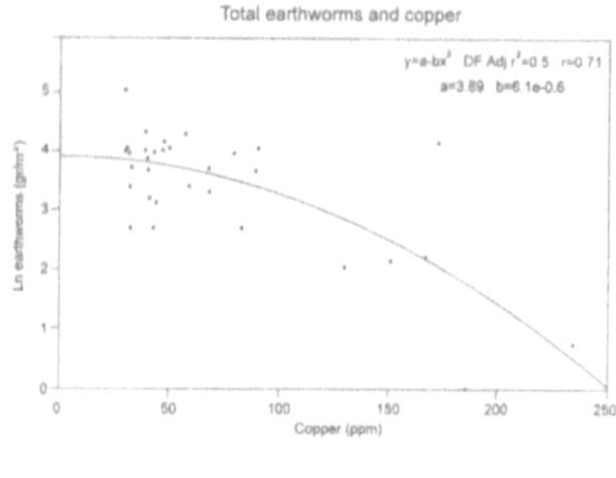

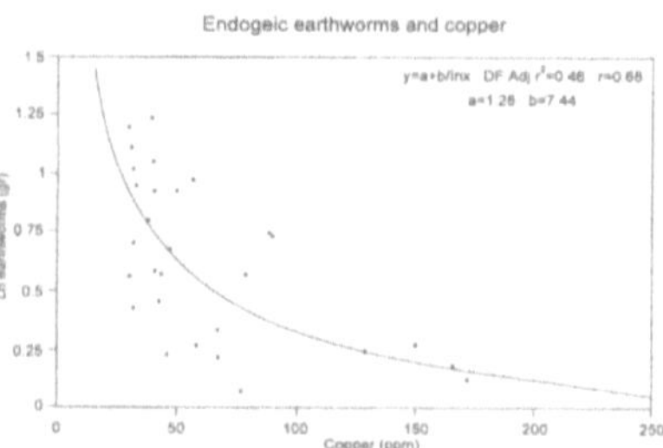

Figure 9, 10. Copper and zinc in different orchards. Each value is the mean of 16 samplings. Differences between orchards: ANOVA test: copper $p < 0.0001$; zinc $p = 0.06$.

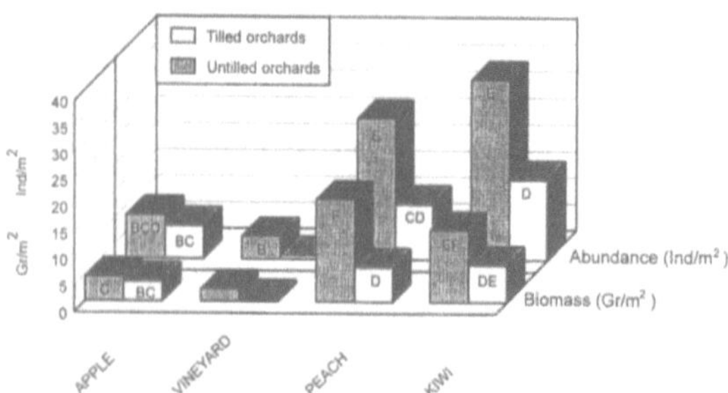

Figure 11. *Aporrectodea caliginosa* Sav. Biomass and abundance in different orchards; tillage differences: $p = 0.001$ for biomass and $p = 0.008$ for abundance; difference in crop plant affect: $p < 0.0001$ for biomass and abundance (Kruskall-Wallis test).

Aporrectodea caliginosa as well as all earthworms were affected by both tillage and chemical treatments. Tillage reduced abundance and biomass by 50-80% in all orchards apart from apple orchards (16% reduction for biomass and 24% for abundance). This species almost disappeared in tilled vineyards (0.85 n/m^2, 0.25 g/m^2) (Fig. 11). It had a negative correlation with copper; for biomass $\ln y = a - b \times 1.5$, $r^2 = 0.19$, $r = 0.44$, $p < 0.01$; for abundance $\ln y = a - b \times 0.5$, $r^2 = 0.17$, $r = 0.37$, $p < 0.05$.

Allolobophora chlorotica was highly negatively affected by copper treatment, whereas tillage produced no effect for this earthworm's abundance (ANOVA test, $p = 0.36$). Biomass was higher in tilled orchards then in untilled ones, even if ANOVA test do not detect such differences as significant ($p = 0.15$). In this case tillage seemed to increase this earthworm's biomass and abundance (Fig. 12). The endogeic group as a whole was negatively affected by both tillage and chemical input. Tillage caused a great reduction of 40-60%, except in apple orchards where a modest, not statistically significant increase occurred in tilled orchards (Fig. 13). Vineyards support the lowest number of those organisms. The negative correlation with copper was highly significant.

These different orchards were easily assessed by using earthworm abundance, species dominance, and diversity. In particular tillage and pesticide residues were the key factors that were possible to assess.

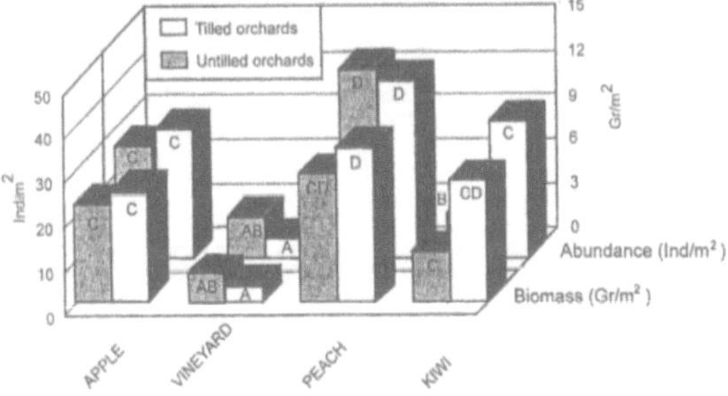

Figure 12. Allolobophora chlorotica Sav. Biomass and abundance in different orchards; tillage differences: $p = 0.036$ for biomass and $p = 0.15$ for abundance; difference in crop plant effect: $p < 0.0001$ for biomass and abundance (Kruskall-Wallis test).

3.4.4 VINEYARDS ASSESSMENT IN SPAIN

As an illustration of the consequences of recent agricultural transformation on arthropod biodiversity in Mediterranean countries, we can show the preliminary results of a study using pitfall trapping carried out in a transformed vineyard area in Alicante Province (SE Spain) (partial results can be found in Perez-Martinez, 1997). Present agricultural landscape in this area is the result of a recent transformation of traditional Mediterranean dry-farming terraced lands (cereal, almond trees, and olive trees) into irrigated vineyards with high chemical input. These modern, intensive farming systems have generated several environmental problems: underground water table exhaustion or salinisation, contamination of waters by fertilizers, etc. Several years after transformation, these high input systems have also been shown to be not economically viable. Thus, many previously transformed fields are now being abandoned, due principally to the high cost of irrigation water, now transported from distant places, and also to marketing problems. EC policy has promoted set-aside as an environmentally sound measure to cope with economical unsustainability. It was the aim of the study to analyze the response of Coleopteran fauna to agricultural intensification and, also, to test if set-aside and reversion of agricultural land to scrub land or forests is really a good strategy for enhancing biodiversity.

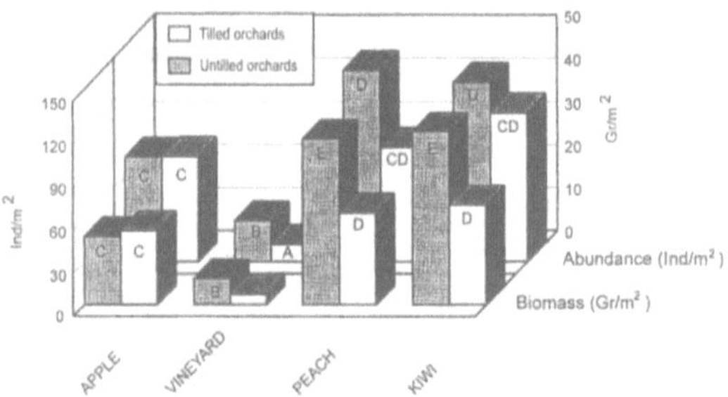

Figure 13. "Endogees" earthworms biomass and abundance in different orchards; tillage differences: $p = 0.007$ for biomass and $p = 0.0007$ for abundance; difference in crop plant effect: $p < 0.0001$ for biomass and abundance (Kruskall-Wallis test).

As shown in Figure 14a, b, terrestrial Coleopteran abundance and species richness are greatly affected by agricultural transformations, total captures of individuals and species were always lower in intensive vineyards than in traditional almond-tree fields.

However, proximity to natural vegetation areas increases, to a certain extent, terrestrial Coleopteran abundance in vineyards, since captures are progressively lower as field distance to natural areas increases (T6 is in direct contact with natural areas, T7 at a some distance, and T8 at a greater distance). Duelli *et al.* (1989) found a similar picture in maize fields in Switzerland. Apparently, natural, uncultivated areas serve as reservoirs of arthropods from which agricultural lands can be colonized. So, impoverishment of Coleopteran fauna in agricultural fields would be greater if cultivated fields extended monotonically over large areas, as is the trend nowadays. Small remnants of natural vegetation were generally present in traditional cultivated areas because of irregularities of land relief, small creeks and hills, etc, which were hard to put into cultivation. Present technology allows the removal of these accidents and the enlarging and leveling of vineyard fields also over ancient natural areas.

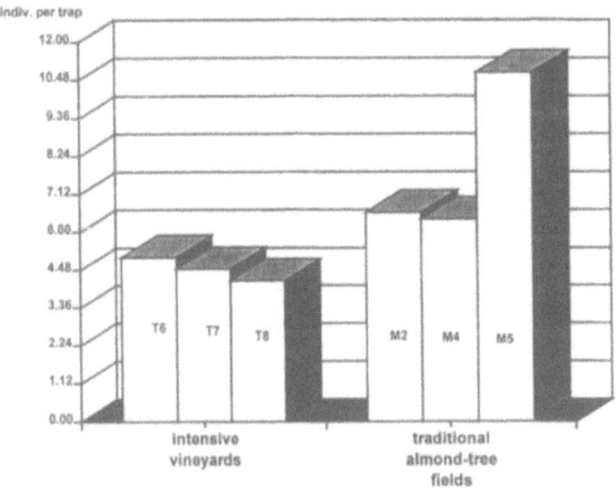

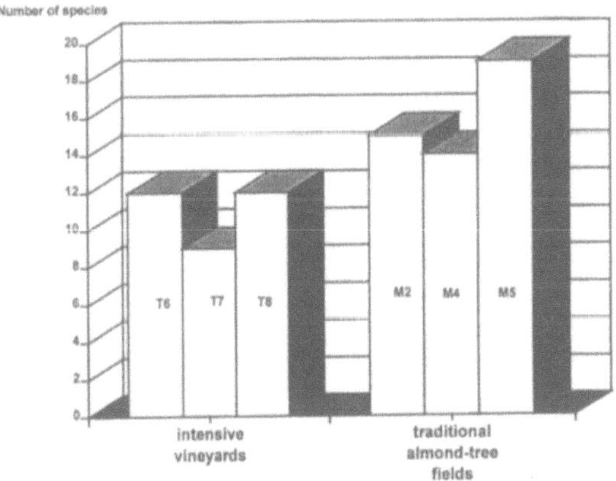

Figure 14a, b. Terrestrial Coleoptera captured by means of pitfall traps in different cultivated fields in Alicante province (SE Spain) during autumn 1997. T6, T7 and T8 are placed in intensive, high-input irrigated vineyards; M2, M3 and M4 in traditional, dry-farming almond-tree terraces. In **a**, total captures per trap; in **b**, total number of species captured in each field. In order to see the influence on vineyard coleoptera fauna of the distance from natural areas, vineyard experimental fields were placed at different distances from an uncultivated hill covered with semi-arid Mediterranean shrub land. T6 was in contact with natural areas, T7 at a middle distance (ca. 100 m.), and T8 at a greater distance (ca. 300 m). All almond-tree fields were at a middle distance from the closest natural area.

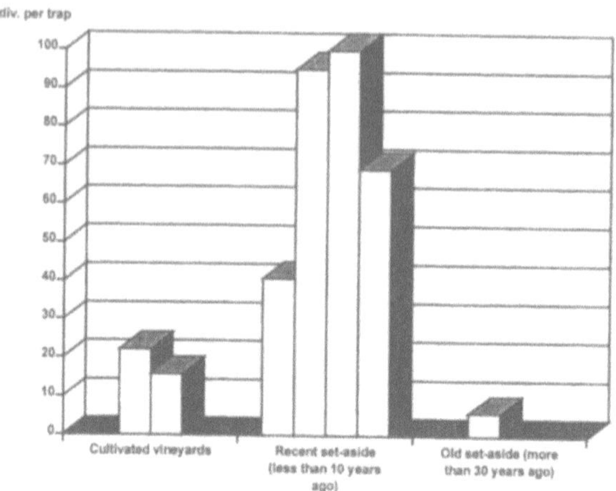

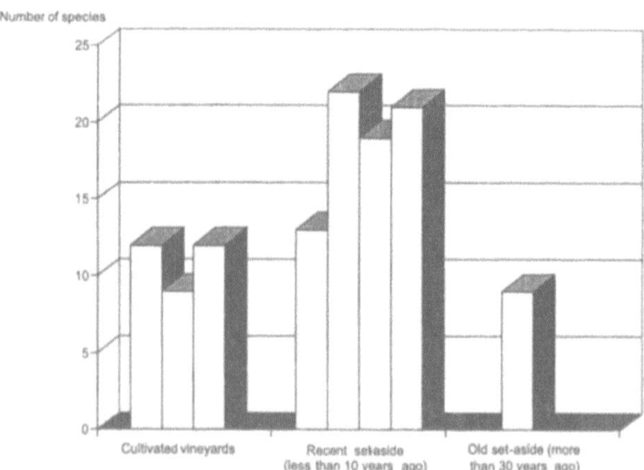

Figure 15a, b. Consequences of set-aside on the Coleoptera fauna of a vineyard area in SE Spain, as indicated by pitfall trapping in cultivated fields and in fields abandoned at different periods (5-8 years ago, with a vegetation cover dominated by ruderal communities, and more than 30 years ago, with well-developed Mediterranean scrub land). In **a:** total number of individuals captured per trap; in **b:** total number of species captured in each field.

With respect to the consequences of set-aside, in Figure 15a, b, the captures per trap and species richness in cultivated vineyards are shown for different short-time set-aside fields (less than 10 years) and a long-time set-aside field (more than 30 years). As a global trend, set-aside favors Coleopteran abundance and species richness during the first years, total captures passing from 3.87-5.12 individuals/trap in cultivated vineyards to 40.62-100.00 individuals/trap in 5-8 year set-aside fields. Although conditions are not strictly similar, T1-T4 fields can be compared, to a certain extent, with ancient traditional fallow fields (fallow periods of 3, even 5 years were not rare in the region). Dates can be illustrative of the importance of the presence of fallow patches (another reservoir of soil fauna) in maintaining higher arthropod richness in the entire agrolandscape, and of the impacts on arthropod fauna of the permanent, non-fallow cultivation systems typical of modern agriculture.

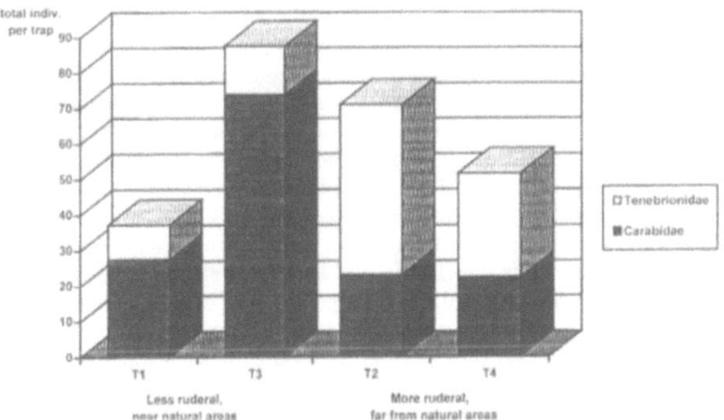

Figure 16. Total captures per trap of Carabidae and Tenebrionidae in set-aside fields in a vineyard area of SE Spain. Dominant vegetation cover was formed by perennial ruderal communities (esp. *Oryzopsis miliacea*). T1 and T3, situated near a natural area, present a higher proportion of scrub land species (*Rosmarinus*, *Thymus*, etc.). T2 and T4, situated at a greater distance from any natural area, represent a higher proportion of ruderal species (including a significant percentage of Chenopodiaceae and Amaranthaceae). T3 was the only field connected to natural areas by a field margin with a dense plant cover.

However, total Coleopteran abundance decreased dramatically in the field abandoned for a long period (more than 30 years) and in a relatively well-developed Mediterranean scrub land (Fig. 15a and b). Here only 6.00 individual/trap and 9 species were captured. It can be deduced that, as succession proceeds and Mediterranean sclerophilous scrub develops, terrestrial Coleoptera, after an initial increase in the first years, become progressively less abundant, reaching even lower values than in intensive vineyards. Similar results were obtained by Schnitter (1994) in Germany.

If we analyze separately the abundances of the two predominant families, Tenebrionidae and Carabidae in the different set-asides, we can have a more detailed picture of the pattern of this agro-landscape system (Fig. 16). Both Carabidae and Tenebrionidae increase their abundances with cessation of cultivation, as was shown in Figure 15a and b. In the case of Tenebrionidae, subfamily Pimeliinae, absent from irrigated vineyards, is now well-represented in set-aside fields. This is likely due to the new dry, no-irrigation conditions. However, apparently Carabidae are dominant over Tenebrionidae in set-aside fields placed closer to natural scrub land areas, where there is a more developed plant cover, undoubtedly due to the proximity of non-ruderal plant propagules (case of T1 and T3). Greater abundance of Coleoptera in T3 than in T1 can be put in relation to the higher connectivity of T3 to natural areas, since T1 is at a greater distance from scrub lands, and additionally, not connected to them by vegetation-covered field margins. On the contrary, Tenebrionidae, a taxon composed by xerophilous specialists, are more abundant in fields far from natural areas which present a more ruderal (earlier successional stage) and scarce vegetation cover (T2 and T4). Subfamily Pimeliinae, which are even greater desert specialists than subfamily Tenebrioninae, also shows a more acute response to the prevalence of dry conditions, and increase their relative abundance over Tenebrioninae in fields T2 and T4. In advanced succesional stages, as represented by field T5, Carabidae are, as expected, more abundant than Tenebrionidae, almost exclusively Tenebrioninae, and almost no members of subfamily Pimeliinae (ratio Tenebr./Pimel. = 7/1). Thus, the relative abundance of Tenebrionidae to Carabidae, and Pimeliinae to Tenebrioninae, seems directly related to the prevalence of arid, poor vegetation cover conditions. Tenebrionidae are favored in the first stages of secondary succession, and Carabidae in more advanced stages when there is a more developed vegetation cover. The value of the ratio Carabidae/Tenebrionidae as bioindicator of arid environmental conditions has also been pointed out by Marcuzzi (1981), de los Santos (1982) and Martin-Cantarino (1994).

It can be deduced from above-mentioned results, than if intensification of agriculture implies a considerable impoverishment of Coleopteran fauna, permanent set-aside at large scale can have the same, or an even worse effect. However, set-aside of small fields, intermingled and well connected with productive lands, would enhance the richness of Coleopteran communities in cultivated fields. An even greater enrichment would be attained with the combination of fallow fields (that is, short-time set-aside of no more than 5-8 years), natural areas, and non-intensive productive fields, *i.e.*, the kind of complex mosaic present in the traditional landscape.

3.5 References

Beavis, I.C., 1988. *Insects and other invertebrates in Classical Antiquity*. Exeter University Press, Exeter.

Brussaard, L., Behan Pelletier, V.M., Bignell, D.E., Brown, V.K., Didden, W., Folgarait, P., Fragoso, C., Freckman, D.W., Gupta, V.V.S.R., Hattori, T., Hawksworth, D.L., Klopatek, C., Lavelle, P., Malloch, D.W., Rusek, J., Soderstrom, B., Tiedje, J.M. & Virginia, R.A., 1997. Biodiversity and Ecosystem Functioning in Soil. *Ambio* **26**:563-570.

BSTID (Board on Science and Technology for International Development) 1996. *Lost Crops of Africa*. National Academic Press, Washington,, DC, 1:1-383.

Carter V.G. & Dale, T., 1974. *Topsoil and Civilization*. University of Oklahoma Press, Norman, Oklahoma, 291 pp.

Conway G.R. & Barbier, E., 1990. *After the Green Revolution: Sustainable Agriculture for Development.*

Earthscan Publishers, London.

De los Santos, A., 1982. Biologia y ecologia de algunas poblaciones de coleopteros terrestres en dos ecosistemas deil Bajo Guadalquivir. Ph.D. Thesis. Universidad de Murcia (unpublished).

Dindal, D., 1989. *Soil Biology Guide*. J.Wiley & Sons, New York.

Duelli, P., Studer, M. & Marchand, I., 1989. The influence of the surroundings on Arthropod diversity in maize fields. *Acta Phytopathologica et Entomologica Hungarica* **24(1-2)**: 73-76.

Heliövaara, K & Väisänen, R., 1993. *Insects and pollution*. CRC Press, Boca Raton, Florida.

Marcuzzi, G., 1981. Aspetti ecologici della Tenebrionidofauna del Mediterraneo. *Ecol. Medit.* **7**: 103-118.

Martin-Cantarino, C., 1994. Ecologia de los coleopteros Tenebrionidae en un ecosistema de dunas costeras de la provincia de Alicante. Ph.D.Thesis. Universidad de Alicante (unpublished).

NRC (National Research Council) 1989. *Lost Crops of the Incas: Little-known plants of the Andes with promise for Worldwide cultivation*. National Academy Press, Washington, DC, 415 pp.

Paoletti, M.G., 1995. Biodiversity, Traditional Landscapes and Agroecosystem Management. *Landscape and Urban Planning* **31**: 117-128.

Paoletti, M.G., 1997. Are there alternatives to wheat and cows in order to improve landscape quality and biodiversity? In: Napier, T. Camboni, S. & Tvrdon, J. (Eds.), *Soil and Water Conservation Polices: Successes and Failures*. Water and Soil Conservation Society press, in press.

Paoletti M.G., 1997a. IPM practices for Reducing fungicide Use in Fruit Crops. In: D. Pimentel, D. (Ed.), *Techniques for Reducing Pesticide Use: Economic and Environmental Benefits*. J. Wiley and Sons Ltd, pp. 343-378.

Paoletti, M.G., 1998. *Biodiversity in Agroecosystems: Role for sustainability and Bioindication*. Elsevier (in press).

Paoletti, M.G., 1999a. Some unorthodox thoughts: what Western Agriculture should learn from Chinese Agriculture. *Critical Review in Plant Sciences* **18(3)**: 475-487.

Paoletti, M.G., (Ed.), 1999b. Invertebrate biodiversity bioindicators of sustainable landscapes. Pratical use of invertebrates to assess sustainable landuse. *Agriculture, Ecosystems & Environment* **74(1-3)**: 1-444.

Paoletti, M.G. & Bukkens, S.G.F., 1997. Minilivestock. *Ecology of Food and Nutrition (s.i.)*, **36 (2-4)**: 95-346.

Paoletti, M.G., Favretto, M.R., Marchiorato, A., Bressan, M. & Babetto, M., 1993. Biodiversità in pescheti forlivesi. In: Paoletti, M.G. (Ed.), *Biodiversità negli Agroecosistemi. Osservatorio Agroambientale*. Centrale Ortofrutticola, Forlì, pp. 20-56.

Paoletti, M.G., Dreon, A.L. & Lorenzoni, G.G., 1995a. Edible Weeds "Pistic" found in W. Friuli, (NE Italy)., *J. Econ. Bot.* **49(1)**: 26-30.

Paoletti, M.G., Schweigl, U. & Favretto, M.R., 1995. Soil Macroinvertebrates, Heavy Metals and Organochlorines in Low and High Input apple Orchards and a Coppiced Woodland. *Pedobiologia* **39**: 20-33.

Paoletti, M.G., Sommaggio, D., Favretto, M.R., Petruzzelli, G., Pezzarossa, B. & Barbafieri, M., 1998. Earthworms as useful Bioindicators of Agroecosystem Sustainability in Different input orchards. *Applied Soil Ecology* **10**: 137-150.

Perez-Martinez, F.J., 1997. The influence of field boundaries and distance from semi-natural areas on the regeneration of set-aside land in a Mediterranean ecosystem: the example of Coleoptera and plant communities. MSc. Dissert. University College of London (unpublished).

Pimentel, D., Harvey, C., Resosudarmo, P., Sinclair, K., Kurz, D., McNair, M., Crist, S., Shpritz, L., Fitton, L., Saffouri, R. & Blair, R., 1995. Environmental and economic costs of soil erosion and conservation benefits. *Science* **267**:1117-1123.

Pimentel, D., (Ed.), 1997. *Techniques for Reducing Pesticide Use: Economic and Environmental Benefits*. J. Wiley and Sons Ltd.

Pimentel, D. & Pimentel, M., 1996. *Food, Energy and Society*. University Press of Colorado, Niwot, Co.

Rackham, O., 1986. *The History of the Countryside*. The Orion Publ. Group, London, 444 pp.

Schnitter, P.H., 1994. The development of carabid communities from uncultivated fields and meadows in the first five years of a succession. In: Desender, K. *et al.* (Eds.), *Carabid beetles: ecology and evolution.* Kluwer Academic Publ., The Hague, pp. 361-366.

Schweigl, U., 1989. Comparazione pedozoologica ed impatto di residui di pesticidi in agroecosistemi sudtirolesi. Tesi di Laurea, Dip. Biologia, Università di Padova.

Van Straalen, N.M., 1997. Community Structure of Soil Arthropods as Bioindicators of soil health. In: Pankhurst, C., Doube, B.M. & Gupta, V.V.S.R. (Eds.), *Biological Indicators of Soil Health.* Cab International, London, pp. 235-263.

CHAPTER 4

EFFECTS OF HABITAT FRAGMENTATION ON PLANT-INSECT COMMUNITIES

ANDREAS KRUESS *and* TEJA TSCHARNTKE
Agroecology, Georg - August University, Göttingen, Germany

4.1 Introduction

Changes in landscape structure due to human activities includes habitat destruction and the fragmentation of the remaining habitat patches (Harris, 1984). This process of habitat fragmentation has been perceived as "the principle threat to most species in the temperate zone" (Wilcove *et al.*, 1986) or "the single greatest threat to biological diversity" (Noss, 1991). Although habitat fragmentation occurs naturally, it is mostly caused by the expansion and intensification of anthropogenic land use (Burgess & Sharpe, 1981). For example, in the Australian wheat belt region 93% of the native vegetation has been cleared, mostly during the last 50 years (Saunders *et al.*, 1993). Estimating the current effects of fragmentation on species diversity is often difficult (Margules *et al.*, 1994), and most investigations have studied the effects *a posteriori* (Villard & Taylor, 1994). As mentioned by Didham *et al.* (1998), little attention has been paid to alteration in the trophic structure of communities due to habitat fragmentation, because most studies have focused on single species or several species within one trophic level. Since habitat fragmentation does not affect all species equally, results from systems with such reduced levels of complexity cannot be extrapolated to explain responses of food-web or community interactions.

The equilibrium theory of island biogeography (MacArthur & Wilson, 1967) predicts species numbers on islands as a function of island area and isolation. With regard to biological conservation it is important to know which kind of species will go extinct first or will be particularly negatively affected by habitat fragmentation. Greatly endangered species show population characteristics like rarity (reduced abundance and distribution), high population variability (enhanced fluctuations), high degrees of specialization, dependence on mutualists, or little dispersal ability (see also Lawton, 1995; Tscharntke & Kruess, 1999).

Fragmentation of habitats is characterized by at least three important processes each affecting the diversity and the spatial distribution of species (Andrén, 1994): (i) area reduction of the original habitat in the landscape due to habitat loss; (ii) area reduction of the emerging habitat fragments; and (iii) increasing distance between the fragments. As a consequence of the reduced island area, edge effects may have additional effects on diversity and species distribution pattern. These three major features of fragmentation processes are related in a non-linear way (Gustavson & Parker, 1992; Andrén, 1994).

B. Ekbom, M. Irwin and Y. Robert (eds.), Interchanges of Insects, 53-70

When fragmentation affects critical proportions of habitat, rapid changes in size and isolation of the habitat fragments can crop up (Turner, 1989; Gustavson & Parker, 1992; Andrén, 1994; Bascompte & Solé, 1996).

Andrén (1994, 1996) and Bascompte & Solé (1996) found from mathematical modeling that for low-level habitat loss the quantitative effects of area reduction is the dominating process. Dissociation of the original habitat into fragments becomes more significant when habitat loss reaches 40%. Further losses of habitat caused rapid increases in the number of habitat fragments. When habitat fragmentation reaches 80%, the number of habitats declined heavily (Bascompte & Solé, 1996), and isolation of the habitat fragments exponentially increased (Andrén, 1994).

In modern agriculture habitat loss on a landscape scale has often reached 80% or more (*e.g.* Saunders *et al.*, 1993). At such a high level of fragmentation, isolation appears to be a major threat to biological diversity. In addition, areas of near-natural habitats still existing in the agricultural landscape often enclose a wide range, from small patches with only a few hundred square meters to large areas extending hundreds of hectares. This pattern does not meet expectations from the results of computer-simulated fragmentation processes mentioned above. An explanation for this may be that real fragmentation processes are not random, but powered by economic decisions due to landscape management or geographic conditions. For example road building causes a more regular (but not random) "cutting pattern" of the landscape. Significance of habitat area depends on the species or taxa under investigation. For example, invertebrates can cope with smaller islands than vertebrates.

Moreover, the effects of fragmentation on a particular species depend on its ecological requirements (*e.g.* home range) or biological characteristics (*e.g.* mobility, body size). For example species with large home ranges like many birds are not affected by habitat isolation on a local scale because their territories may include several habitat patches (Tjernberg *et al.*, 1993). For those species, the consequences of habitat fragmentation are only the quantitative effects of habitat loss, but not the qualitative effects of area or isolation of the remaining habitat islands.

In this chapter, we present some empirical support for answers to the following questions: (1) Does habitat fragmentation negatively affect insect diversity? (2) Are the effects equal for different trophic levels? (3) What are the consequences of fragmentation for herbivore-parasitoid interactions? (4) How can the most affected species be characterized? (5) What are the consequences for biological conservation with respect to the SLOSS debate?

To answer these questions we will focus on two well-known plant-insect communities, comprising both endophagous herbivores and their parasitoids (Kruess & Tscharntke, 1994, 1999; Kruess, 1996; Kruess, 1998). Each of the two insect communities were centered on a single plant species, red clover (*Trifolium pratense*) and bush vetch (*Vicia sepium*).

We investigated the effects of habitat fragmentation in two different ways: (1) we analyzed both insect communities on near-natural old meadows, nearly equal in management regime and vegetation structure, but different in area and isolation (Kruess, 1996; Kruess & Tscharntke 1999). (2) we experimentally analyzed the colonization process of both insect communities on manually established, small and isolated plant plots in the agricultural landscape (Kruess & Tscharntke, 1994; Kruess, 1996).

4.2 Using Plant-Insect Communities as Model Ecosystems

Since the studied insect communities comprise both herbivores and parasitoids, changes in species diversity can be analyzed on different trophic levels. Moreover, effects on interactions among these trophic levels can be determined. Most studies on the effects of fragmentation on plant-insect systems have analyzed ectophagous insect communities (*e.g.* on nettle by Davis, 1975; Zabel & Tscharntke, 1998; on juniper by Ward & Lakhani, 1977; on bracken by Rigby & Lawton, 1981). Only few studies have included investigations on endophagous insects (MacGarvin, 1982; Davis & Jones, 1986). Endophagous insect communities are more likely to comprise high proportions of monophagous herbivores and parasitoids, so that isolation or habitat area can be easily defined. Polyphagous species are less sensitive to fragmentation processes than monophagous species (Zabel & Tscharntke, 1998).The inclusion of generalists in analyses of species composition of habitat islands can lead to an overestimation of species diversity in small habitats because species depending on surrounding habitats in the landscape are also included (Loman & von Schantz, 1991). For generalists, populations on habitat islands are not isolated, but closely connected with conspecific populations in the surrounding landscape.

We focused on the endophagous insects in the flowerheads and stems of red clover (*Trifolium pratense*), and in the pods of vetch (*Vicia sepium*), comprising mostly mono- or oligophagous herbivores and their parasitoids. The two plant species are abundant and typical representatives of the investigated meadows.

Most studies that have included more than one trophic level of insect communities have analyzed herbivores and predators (Davis, 1975; Ward & Lakhani, 1977; Kareiva, 1987; Spiller & Schoener, 1988; Zabel & Tscharntke, 1998). Since predators are on average less specialized than parasitoids, investigations of multitrophic interactions based on host-parasitoid associations are more likely to show island effects. In the following we will briefly describe the two insect communities.

4.2.1 ENDOPHAGOUS INSECTS ON RED CLOVER (TRIFOLIUM PRATENSE)

Dissections of flower heads and stems of red clover revealed an insect community of 23 species, comprising 8 herbivores and 15 parasitoids (Table 1). The most abundant herbivorous species were the seed-feeding weevil *Protapion apricans*, the two stem-boring weevils *Catapion seniculus* and *Ischnopterapion virens*, the seed beetle *Bruchidius varius*, and the seed-feeding chalcid wasp *Bruchophagus gibbus*. Parasitoids were associated only with the weevils of the family Apionidae and an undescribed gall midge species, *Lasioptera* sp. nov. The most abundant parasitoid species were the pteromalid wasps *Spintherus dubius, Trichomalus campestris, T. fulvipes*, two unidentified eulophid wasps of the genus *Aprostocetus*, and two braconid wasps of the genus *Triaspis*. All but one species (*Spintherus dubius* with 79% of all specimens) were relatively rare (less than 10% of all specimens). All but one of the herbivores feed on red clover only. The weevil *Protapion assimile* also feeds on white clover (*Trifolium repens*), a clover species occurring on only few of such meadows and in small populations. As far as we know, all but five of the parasitoids attack only hosts on red clover, but four species (*Triaspis obscurellus, Spintherus dubius, Stenomalina gracilis, Trichomalus campestris*) were also known to

Table 1. Endophagous insect community in the flower heads and stems of *Trifolium pratense*. Site of attack of the herbivores and host genus of the parasitoids are listed in the third column, host range of both herbivores and parasitoids is given in the right column (Oligophag = hosts within one family, Polyphag = hosts from more than one family, ? = unknown)

Herbivores	**Family**	**Site of attack**	**Host range**
Catapion seniculus (Kirby)	Apionidae (Col.)	stems	Monophag
Ischnopterapion virens (Herbst)	Apionidae (Col.)	stems	Monophag
Protapion apricans Herbst	Apionidae (Col.)	flower heads	Monophag
Protapion assimile (Kirby)	Apionidae (Col.)	flower heads	Oligophag
Protapion trifolii (Linnaeus)	Apionidae (Col.)	flower heads	Monophag
Bruchidius varius Olberg	Bruchidae (Col.)	flower heads	Monophag (oligophag?)
Bruchophagus gibbus Boheman	Eurytomidae (Hym.)	flower heads	Monophag (oligophag?)
Lasioptera sp. nov.	Cecidomyiidae (Dipt.)	stems	Monophag?
Parasitoids	**Family**	**Host genus**	
Colastes sp.	Braconidae (Hym.)	*Catapion, Ischnopterapion*	?
Triaspis obscurellus (Nees)	Braconidae (Hym.)	*Catapion, Ischnopterapion, Protapion*	Oligophag
Triaspis floricola (Wesmael)	Braconidae (Hym.)	*Catapion, Ischnopterapion*	Monophag?
Spintherus dubius (Nees)	Pteromalidae (Hym.)	*Protapion*	Oligophag
Stenomalina gracilis (Walker)	Pteromalidae (Hym.)	*Catapion, Ischnopterapion*	Oligophag
Trichomalus campestris (Walker)	Pteromalidae (Hym.)	*Catapion, Ischnopterapion, Protapion*	Oligophag
Trichomalus fulvipes (Walker)	Pteromalidae (Hym.)	*Catapion, Ischnopterapion*	Monophag
Trichomalus helvipes (Walker)	Pteromalidae (Hym.)	*Protapion*	Monophag
Aprostocetus cf. *tompanus* Erdös	Eulophidae (Hym.)	*Catapion, Ischnopterapion*	Monophag?
Aprostocetus vassolensis Graham	Eulophidae (Hym.)	*Lasioptera*	Monophag?
Aprostocetus sp. 1	Eulophidae (Hym.)	*Protapion*	?
Aprostocetus sp. 2	Eulophidae (Hym.)	*Protapion*	?
Entedon cf. *procioni* Erdös	Eulophidae (Hym.)	*Catapion, Ischnopterapion*	Oligophag
Pseudotorymus apionis (Mayr)	Torymidae (Hym.)	*Protapion*	Monophag
Eupelmus vesicularis (Retzius)	Eupelmidae (Hym.)	*Catapion, Ischnopterapion,* and others	Polyphag

Table 2. Endophagous insect community in the pods of *Vicia sepium*. Site of attack of the herbivores and host genus of the parasitoids are listed in the third column, host range of both herbivores and parasitoids is given in the right column (Oligophag = hosts within one family, Polyphag = hosts from more than one family, ? = unknown)

Herbivores	**Family**	**Site of attack**	**Host range**
Oxystoma ochropus (Germar)	Apionidae (Col.)	pods	Monophag (oligophag?)
Tychius quinquepunctatus (L.)	Curculionidae (Col.)	pods	Monophag (oligophag)?
Bruchus atomarius (L.)	Bruchidae (Col.)	pods	Oligophag
Cydia nigricana (F.)	Tortricidae (Lep.)	pods	Oligophag
Parasitoids	**Family**	**Host genus**	
Pristomerus vulnerator (Pz.)	Ichneumonidae (Hym.)	*C. nigricana*, and other moths	Polyphag
Scambus annulatus (Kiss)	Ichneumonidae (Hym.)	*C. nigricana*	Oligophag?
Pigeria piger (Wesmael)	Braconidae (Hym.)	*C. nigricana*	Monophag
Glabrobracon sp.1	Braconidae (Hym.)	*C. nigricana*, and other moths?	?
Glabrobracon sp.2	Braconidae (Hym.)	*C. nigricana*, and other moths?	?
Triaspis thoracicus (Curt)	Braconidae (Hym.)	*B. atomarius*	Oligophag
Pteromalus sequester (Walker)	Pteromalidae (Hym.)	*O. ochropus*	Monophag?
Trichomalus repandus (Walker)	Pteromalidae (Hym.)	*O. ochropus*	Monophag?
Entedon cf. *procioni* Erdös	Eulophidae (Hym.)	*O. ochropus*, and others	Oligophag
Eupelmus vesicularis (Retzius)	Eupelmidae (Hym.)	*O. ochropus*, and others (see tab.1)	Polyphag

feed on weevils on white clover, and the eulophid wasp *Entedon* cf. *procioni* was also found attacking a weevil in the pods of the vetch *Vicia sepium* (see Table 2). The only known polyphagous parasitoid was the eupelmid wasp *Eupelmus vesicularis*.

4.2.2 ENDOPHAGOUS INSECTS ON THE VETCH (VICIA SEPIUM)

In the pods of the vetch we found an insect community of 14 insect species, consisting of 4 herbivores and 10 parasitoids. The two weevils *Tychius quinquepunctatus* and *Oxystoma ochropus* are known to feed also in the pods of other *Vicia* and *Lathyrus* species (Dieckmann, 1977; Freude *et al.*, 1981), but this was not supported by our rearings. The seed beetle *Bruchus atomarius* is also known to feed on *V. villosa*, a plant species that is occasionally found in crop fields or on ruderal habitats. The moth *Cydia nigricana* is a well-known herbivore attacking *Lathyrus, Pisum,* and *Vicia* species. The braconid wasp *Triaspis thoracicus* is oligophagous on bruchid beetles; the two pteromalid wasps *Pteromalus sequester* and *Trichomalus repandus* are mono- or oligophagous on apionid weevils in pods of *Vicia* or *Lathyrus* species. *Entedon* cf. *procioni* is oligophagous on apionid weevils, *Pristomerus vulnerator* is polyphagous on Microlepidoptera, and *Eupelmus vesicularis* is polyphagous on various hosts. The status of the other species is more or less unknown.

4.2.3 RESEARCH AREA

The investigation was carried out between 1992 and 1994 in south-west Germany near Karlsruhe. The study area, the "Kraichgau", located 20 km north-east of Karlsruhe, is a greatly diversified agricultural landscape with a heterogeneous mixture of agricultural fields, hedges, woodland, and meadows.

The meadows of this region are a typical example of formerly widely distributed habitats spanning areas of more than 100 ha. Due to intensification of human activities (urban development, modern agriculture) since the 1950's, both the number of large meadows and meadow area have declined (Hölzinger, 1987). However, there is still a pattern of a few large meadows with more than 10 ha and many patchily distributed small meadow fragments, spread over the landscape and surrounded by agricultural areas. The meadows are sparsely planted with apple and cherry trees and differ in mowing intensity from once to three times a year (we used only those meadows that were mown once a year, in July). Habitat characteristics and vegetation structure were analyzed on 27 meadows, ranging in size from 0.03 to 70 ha. Finally, only 20 meadows were used to analyze the insect communities, since the others differed too much in vegetation composition.

4.3 Species richness

The classical species-area relationships from island biogeography theory (MacArthur & Wilson, 1967; Wilson & Willis, 1975; Wilcox, 1980; Wilcox & Murphy, 1985; Diamond & May, 1981) show a positive correlation between habitat size and species richness.

Similar relationships have been found on terrestrial habitats for plant species (Williams, 1964) and many groups of animals (Brown, 1971; Thornton *et al.*, 1993). Species-area relationships can be explained by two hypotheses:

a) The area-per-se hypothesis: assumes that it is only the available area that affects species richness. The mechanisms are: (i) a higher random extinction probability due to smaller population sizes in smaller areas (Shaffer, 1981; Gilpin & Soulé, 1986; Have, 1993; Baur & Erhardt, 1995); (ii) the random sample hypothesis (Connor & McCoy, 1979; Haila, 1983): small habitat fragments are only randomly taken (sub)-samples from the original habitat. The likelihood of finding a certain species will be lower in small habitats, since it only depends on sample-size stochastics. In landscapes with high proportions of habitat and thereby, few fragmentation effects, the random sample hypothesis appears to predict species diversity well (Haila, 1983; Haila & Järvinen, 1983), but in habitats with high levels of fragmentation species diversity is often lower than predicted by the random sample hypothesis (Martin & Lepart, 1989). This may be due to critical thresholds in minimum area requirements or habitat connectivity, where small changes in the spatial pattern produces steep shifts in ecological processes (Kareiva & Wennergren, 1995; With & Christ, 1995; Andrén, 1996; Bascompte & Solé, 1996).
b) The habitat-heterogeneity hypothesis: As habitat area increases it is more likely that different types of habitat are included. Because each habitat type may support different species, the total number of species will be higher in more heterogeneous habitats (Williams, 1964; Johnson & Simberloff, 1974; Begon *et al.*, 1995). Since area and heterogeneity are often very closely related it may be difficult to determine whether area or heterogeneity affects species richness more (Rosenzweig, 1995).

In most cases, species-area relationships are best fitted by log-log regression lines (Begon *et al.*, 1995; Rosenzweig, 1995). The non-linear regression model $S = c A^z$, (S = number of species, A = habitat area, c = intercept, and z = slope) typically results in saturation curves (MacArthur & Wilson, 1967). Accordingly, the most dramatic changes in species richness are to be expected for area loss in small habitats. Species extinction probability in small habitats may be somewhat compensated by the "rescue-effect" of conspecific immigrants (Brown & Kodric-Brown, 1977), as long as the fragments are not negatively affected by isolation.

The values of both c and z vary depending on habitat type, isolation, and type of the species involved. Species-area curves usually have lower z-values in terrestrial habitats than in island habitats, and non-isolated terrestrial habitats generally have lower z-values (0.13 to 0.18) than isolated terrestrial habitats (0.25 to 0.33) (Rosenzweig, 1995). A higher z-value indicates a stronger decline in species richness due to area reduction. Zabel & Tscharntke (1998) found higher z-values for monophagous herbivores compared to polyphagous herbivores. Since species richness on small and isolated habitats depend primarily on immigration processes, species-area curves of those habitats are characterized by low c-values.

Species-area curves of the insect communities that we investigated on meadows differed according to trophic level: the species-area curves of the parasitoids on both plant species, *Trifolium pratense* and *Vicia sepium*, were characterized by significantly higher z-values than the curves of their herbivorous hosts when we used the regression model log S = log c + z log A (z-values: *T. pratense*: herbivores = 0.008, parasitoids = 0.16, $F = 21.9$, $p < 0.001$; *V. sepium*: herbivores = 0.05, parasitoids = 0.17, $F = 7.0$, $p = 0.01$). Thus, parasitoids were more sensitive to habitat loss than their hosts.

Our data from the meadows were not best-fitted by the log-log model but by the model: S = c + z lnA. However, species-area curves based on the regression model S = c + z lnA, shown in Figure 1, also showed these differences between herbivores and parasitoids. Species number of parasitoids declined significantly steeper with habitat loss (from approximately 11 species on the largest meadows to 3 species on the smallest meadows, see Fig. 1b) than the number of their herbivorous host species (from 7 species in the largest to 4 species in the smallest meadows, see Fig. 1a). Each of the meadows smaller than 1 ha supported less than 50% of the total number of parasitoid species.

Species-area relationships of the insect community on the vetch *V. sepium* gave similar results. The slopes of the species-area curves of herbivores and parasitoids differed significantly (Fig. 1c,d). Decline in species richness due to area reduction was steeper for parasitoids (from approximately 5 species in the largest meadows to 1 species in the smallest meadows) than for herbivores (from 4 species in the largest to 3 species in the smallest meadows).

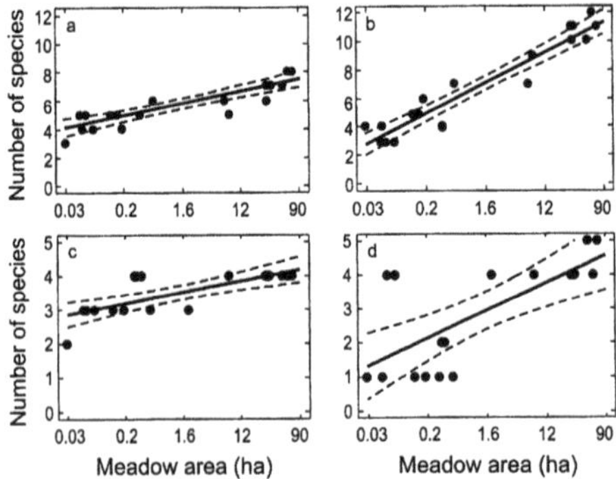

Figure 1. Effect of habitat area on species richness. Endophagous insects on naturally occurring red clover (*Trifolium pratense*) and the vetch (*Vicia sepium*) on differentially sized old meadows: *Trifolium pratense*: a) Herbivores: Y = 1.49 + 0.43*ln X, $F = 59.8$, $r^2 = 0.78$, $p < 0.001$, $n = 19$; b) Parasitoids: Y = -3.35 + 1.07*ln X, $F = 180.1$, $r^2 = 0.96$, $p < 0.001$, $n = 19$; slopes of herbivores and parasitoids differ significantly ($F = 44.3$, $p < 0.001$); *Vicia sepium*: c) Herbivores: Y = 1.91 + 0.17*ln X, $F = 20.4$, r2 = 0.56, $p < 0.001$, $n = 18$; d) Parasitoids: Y = -1.01 + 0.41*ln X, $F = 17.7$, $r^2 = 0.53$, $p < 0.001$, $n = 18$; slopes of herbivores and parasitoids differ significantly ($F = 5.4$, $p = 0.03$). Data from a field study on old meadows in the agricultural landscape of Southwest Germany (Kruess 1996, Kruess & Tscharntke 1999).

Species richness of insects was not the only variable affected by habitat reduction. In our field experiments on the colonization process on manually established, small and isolated plant plots ("plant islands"), we found a comparable pattern due to habitat isolation (Kruess & Tscharntke, 1994; Kruess & Tscharntke 1999). Colonization success and reproduction on clover plots was significantly higher for herbivores than for parasitoids (Kruess & Tscharntke, 1994). Since the estimation of colonization success was based on plant dissections and rearing larvae and pupae, reproduction success could be easily measured. On the most isolated clover plots, only 2 to 4 out of 12 parasitoid species but 6 out of 8 herbivores reproduced successfully. The most abundant parasitoid species, *Spintherus dubius* (79% of the specimens), was the only one found on all 18 clover plots. None of the other parasitoid species provided more than 7% of the specimens. The occurrence of the polyphagous but scarce species *Eupelmus vesicularis* even on the most isolated clover plots may be due to existence of alternate hosts in the surrounding landscape.

Investigations on the colonization success of the insect community on the vetch *V. sepium* underlined the results gained from analysis of the clover insects: species diversity of parasitoids attacking *Vicia* herbivores showed a steeper decline than their hosts due to the isolation of small plant plots (Kruess, 1996; Kruess & Tscharntke, 1999).

The species diversity pattern of insects centred on *T. pratense* and *V. sepium* not only supported the predictions from island biogeography theory (MacArthur & Wilson, 1967; Diamond & May, 1981) but also highlighted that species diversity on higher trophic levels (parasitoids) within insect communities were much more affected by habitat fragmentation processes than those on lower trophic levels (herbivores). Such shifts in species composition, along with changes in species abundance (see below), can negatively affect species interactions. This is discussed in detail later.

On a landscape scale, species diversity in each of a set of habitats is affected by the distribution pattern of the species. Patterns predicted by the random sample hypothesis (Connor & McCoy, 1979) have been found in less fragmented landscapes (*e.g.* Haila, 1983; Haila & Järvinen, 1983) whereas species richness was lower than expected by random sample theory in more fragmented landscapes (Martin & Lepart, 1989). This pattern may indicate the existence of critical thresholds for habitat connectivity (Kareiva & Wennergren, 1995; Kimberly & Christ, 1995; With & Christ, 1995; Andrén, 1996; Bascompte & Solé, 1996). Below these thresholds carrying capacity of the habitats is not reached due to the resulting high extinction rates (Gilpin, 1987; Andrén, 1994).

4.4 Parasitism Rate

As a consequence of the observed differences in sensitivity to habitat fragmentation between herbivores and parasitoids, we might expect a shift in herbivore-parasitoid interactions. Percent parasitism of herbivores is a good estimator for the outcome of classical biocontrol (Hawkins & Gross, 1992; Hawkins *et al.*, 1993). Variation in percent parasitism directly reflects changes in host-parasitoid interactions due to a) species deletions and b) changes in species abundance.

Figure 2 shows the effects of habitat-area reduction and habitat isolation on percent parasitism of the stem-boring weevils *Catapion seniculus* and *Ischnopterapion virens* on red clover (Fig. 2a,b) and the seed-feeding weevil *Oxystoma ochropus* (Fig. 2c,d). We show the results for these species since they were very abundant and attacked by several parasitoid species (Table 1, 2). Parasitism rate of the two stem-boring weevils were pooled, since reared parasitoids could not be exactly assigned to each of the two species.

Percent parasitism of the stem borers declined from approximately 85% in the largest meadows to only 40% in the smallest meadows due to area loss (Fig. 2a). The negative effect of habitat isolation on stem borer parasitization was stronger: percent parasitism declined from approximately 85% in the non-isolated clover plots to 25% in the most isolated plots (Fig. 2b, Kruess & Tscharntke, 1994).

The effects of habitat fragmentation on parasitism of the most abundant insect herbivore in the pods of *V. sepium,* the seed-feeding weevil *Oxystoma ochropus,* were similar (Kruess, 1996; Kruess & Tscharntke, 1999). Percent parasitism declined from 80% in the largest meadows to 40% in the smallest meadows due to area loss (Fig. 2c). Habitat isolation analyzed on small and isolated plots caused a much more dramatic decrease in the parasitism of *O. ochropus*: parasitism rate declined from 80% in non-isolated vetch plots to zero in plots isolated by more than 100 m, since parasitoids totally failed to colonize vetch plots isolated by more than 100 m (Fig. 2d).

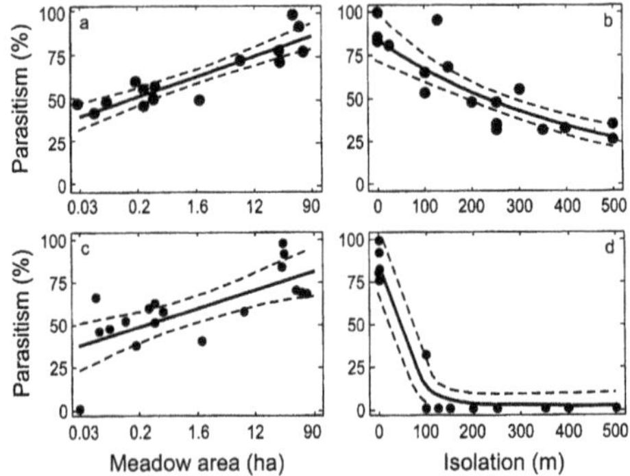

Figure 2. Effects of area and isolation on the parasitism rate of two stem-boring weevils on *T. pratense* (a,b) and a seed-feeding weevil on *V. sepium* (c, d): a) area effect on stem borer parasitism: Y = 14.6 + 2.9*ln X, F = 66.7, $r^2 = 0.82$, $p < 0.001$, n = 19; b) isolation effect on stem borers parasitism: $Y = e^{4.43-0.002*X}$, F = 54.6, $r^2 = 0.77$, $p < 0.001$, n = 18 (reprinted with permission from Kruess & Tscharntke 1994. Copyright 1994 American Association for the Advancement of Science); c) area effect on seed feeder parasitism: Y = 56.0 + 5.6*ln X, F = 16.8, $r^2 = 0.51$, $p < 0.001$, n = 18; d) isolation effect on seed feeder parasitism: Y = 85.3 -15.1*ln (X+1), F = 250, $r^2 = 0.95$, $p < 0.001$, n = 16. Data in a) and c) are from a field study with naturally occurring *Trifolium pratense* on 19 old meadows and naturally occurring *Vicia sepium* on 18 old meadows in the agricultural landscape of Southwest Germany (Kruess, 1996; Kruess & Tscharntke, 1999). Data in b) and d) are from field experiments with manually established small plant plots (18 plots planted with *T. pratense,* 16 plots planted with *V. sepium*), each covering an area of 1.2 m2. The plots differed in isolation, *i.e.* in distance to the nearest old meadow with naturally occurring *T. pratense* and *V. sepium*, and included 5 plots on such meadows as a control (Kruess & Tscharntke 1994, Kruess 1996, Kruess & Tscharntke, 1999). In this figure the maximum parasitism value reached is set to 100 %.

We found that population density of the parasitoids was negatively affected by both habitat area reduction and habitat isolation: Total abundance of parasitoids (i) on red clover attacking the stem-boring weevils *Catapion seniculus* and *Ischnopterapion virens*, and (ii) on the vetch attacking the seed-feeding weevil *Oxystoma ochropus* were positively correlated with both habitat size and host abundance in a stepwise multiple regression (Kruess, 1996). In addition, parasitoid abundance in small plant plots was negatively affected by increasing isolation of the plots. In red clover plots, total abundance of parasitoids attacking stem-boring and the seed-feeding weevils also declined significantly with increasing isolation of clover plots (Kruess & Tscharntke, 1994).

Thus, our investigations of the two herbivore-parasitoid insect communities showed that both features of habitat fragmentation, area loss and habitat isolation, may dramatically disturb herbivore-parasitoid interactions, since parasitoids were more affected than their herbivorous hosts. In consequence, herbivores were greatly released from parasitism. This may lead to higher population densities of herbivores and favor pest outbreaks (Kruess & Tscharntke, 1994). Kareiva (1987) found that population explosions of aphids are more frequent in small and isolated patches of goldenrod.

4.5 Extinction Risks: Species Abundance and Population Variability

In addition to the tropic-level position local abundance of species was related to extinctions: Percent absence of the species on both old meadows and isolated small plant plots was negatively correlated with species abundance (Kruess & Tscharntke, 1994; Kruess, 1996; Kruess & Tscharntke, 1999): Figure 3a shows the negative correlation between percent absence on small plant plots for all insects on *Trifolium pratense*, and Figure 3b shows the same correlation for the insects in the pods of *Vicia sepium*, analyzed on old meadows. Thus, locally rare species are more prone to extinction than very abundant species since small population size enhances the risk of stochastic events that can lead to extinction (Hurka, 1984; Schaal & Leverich, 1984; Soulé, 1987; Pimm, 1991).

As already mentioned above, small habitats support only small populations, and smaller population sizes are linked to higher extinction risks. Local extinction may be compensated by re-colonization due to immigration (Taylor, 1990; Hanski *et al.* 1994), but regular extinctions may also prevent populations from reaching carrying capacity, thereby causing low population densities in small and isolated habitats (MacGarvin, 1982; Andrén, 1994; Hanski *et al.*, 1994; Zabel & Tscharntke, 1998). Since colonization success depends on only a few individuals, reproduction success may be limited in such systems, also leading to lower population densities in small habitats, especially since colonization or immigration of insect species is negatively affected by habitat isolation (Davis, 1975; Ward & Lakhani, 1975; Davis & Jones, 1986; Zabel & Tscharntke, 1998). In contrast, some studies found higher population densities on small and isolated habitat patches compared to large habitats (Kareiva, 1987, 1990; Ferrari *et al.*, 1997). For herbivorous specialists this may be an effect of reduced mortality due to the enhanced extinction probability of natural enemies on small islands (Kareiva, 1990; Kruess & Tscharntke, 1994).

Populations can be characterized by both temporal population fluctuations in each of the habitats and a spatial variability between all populations. Species characterized by stable populations (little fluctuations in time) should also show little spatial variability (similar abundance in different habitats). Species abundance and population variability may be positively correlated (Pimm, 1991; Lawton, 1995), but we found that spatial variability was negatively correlated with local species abundance (in both field experiments with small plant plots and investigations on old meadows for both plant-insect systems). Figure 3c shows the correlation between species abundance and population variability for the insect community in small red clover plots (Kruess & Tscharntke, 1994). Extinction probability was significantly higher for species with highly fluctuating populations. Accordingly, species that failed to colonize small and isolated clover plots were characterized by both low abundance and high population variability (Fig. 3c). For the insect species in the pods of *V. sepium* in 18 old meadows, spatial variability was also negatively correlated with local abundance (Fig. 3d) Thus, species that were prone to extinction were subjected to triple jeopardy: (i) small populations, (ii) high population variabilities, and (iii) high tropic-level position.

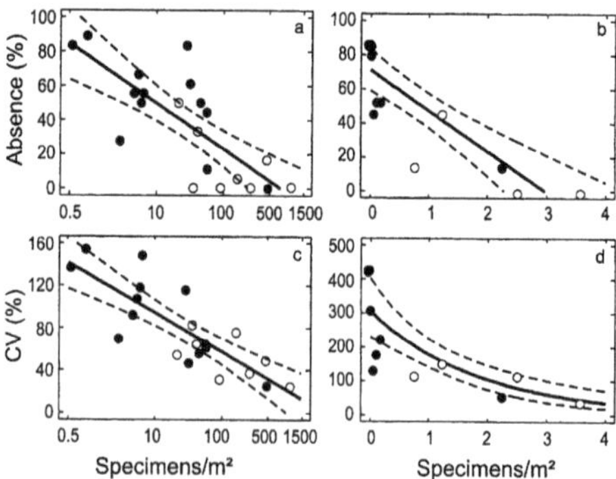

Figure 3. Percent absence and population variability of endophagous insects on small plots of *Trifolium pratense* and on old meadows with naturally occurring *Vicia sepium* (° herbivores, • parasitoids): a) correlation between the absence rate of each of the insects on 18 small plots with *T. pratense* and its average abundance on the 5 control plots: $Y = 77.43 - 11.66 * \ln X$; $F = 28.35$, $r^2 = 0.60$, $p < 0.001$, $n = 21$ (reprinted with permission from Kruess & Tscharntke 1994. Copyright 1994 American Association for the Advancement of Science); b) correlation between the absence rate and the average abundance of each of the insects on *V. sepium* on 18 old meadows: $Y = 71.3 - 0.24\,X$, $F = 35.7$, $r^2 = 0.75$, $p < 0.001$, $n = 14$ (Kruess 1996); c) correlation between the spatial variability (coefficient of variation, CV) of each of the insects on 18 small plots with *T. pratense* and its average abundance on the 5 control plots: $Y = -132 - 0.16 * \ln X$; $F = 38.7$, $r^2 = 0.67$, $p < 0.001$, $n = 21$ (reprinted with permission from Kruess & Tscharntke 1994. Copyright 1994 American Association for the Advancement of Science); d) correlation between the spatial variability (coefficient of variation, CV) and the average abundance of each of the insects on *V. sepium* on 18 old meadow: $Y = e^{5.7-0.005X}$, $F = 34.5$, $r^2 = 0.74$, $p < 0.001$, $n = 14$ (Kruess 1996, 1998).

4.6 SLOSS Debate

Most studies support the idea that both large habitat areas and connectivity of habitats increase species diversity and population stability (Simberloff, 1988; Thomas & Harrison, 1992; Thomas *et al.*, 1992). For the conservation of species diversity on a landscape scale, it is also important to know whether a set of many small or a few large habitats with equal area support the higher species diversity. This leads to the pointed question "single large or several small" (the "SLOSS" debate, see Brown,1986; Quinn & Harrison, 1988; Burkey, 1989). Species-area relationships are not suitable to answer these questions concerning metapopulation diversity because they do not consider species distribution over a set of habitats. In an analysis of data from the literature for oceanic and terrestrial islands, Quinn & Harrison (1988) found that in most cases a set of small islands provides more species than a single large island.

Burkey (1989) argued that the aim to minimize extinction will favor the "single large" strategy, while maximizing species richness will favor the "several small" strategy.

Cumulative species-area curves as used by Quinn & Harrison (1988) showed for both insect communities that species diversity of parasitoids was higher on "several small" habitats. This is shown in Figure 4a for the 14 parasitoid species (= 100%) found on red clover in 19 old meadows. The dotted line is a species-area curve cumulating both species and area, starting with the smallest meadow (0.03 ha) and ending with the largest meadow (70 ha). The solid line cumulates in the reverse order (from the largest to the smallest meadow). In Figure 4b, cumulative species-area curves are plotted for the 10 parasitoid species (= 100%) in the pods of *Vicia sepium* found in 18 old meadows. Since the solid line runs below the dotted line in both cases, species diversity in one or a few large meadows was lower than in a set of several small meadows with the same total area. For example, Figure 4a shows that the minimum area needed to support 90% of the total parasitoid diversity on red clover is only 66 ha if this area is cumulated by

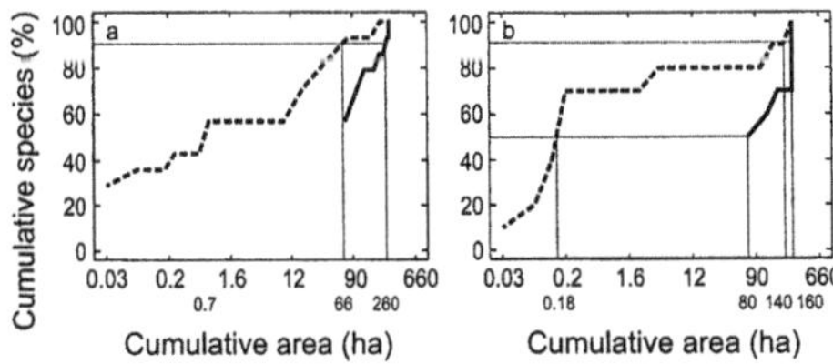

Figure 4. Cumulative species-area curves of parasitoid insect communities on *Trifolium pratense* and *Vicia sepium*: a) cumulation curves of parasitoids on *T. pratense* (14 species on 19 meadows); b) cumulation curves of parasitoids on *V. sepium* (10 species on 18 meadows). Species number and area of the meadows were stepwise cumulated in two contrasting ways: i) starting with the smallest meadow, stepwise adding the smallest but one (dotted lines); ii) starting with the largest meadow, stepwise adding the second largest one (solid lines). The fine-dotted horizontal lines indicate the 50 % and 90 % species diversity. The intersection points with the cumulation curves will give the minimum area needed to support a species diversity of 50 % or 90 %, respectively (for a review see Quinn & Harrison, 1988).

several small habitats (dotted line) but 260 ha if this area is cumulated by a few large habitats (solid line). For the parasitoids on the vetch (Fig. 4b) the difference between the two curves is small on the 90% diversity level (140 ha vs. 160 ha) but great on the 50% diversity level (0.18 ha of several small habitats vs. 80 ha of a few large habitats).

These results, found in both parasitoid communities, can be explained by the low nested distribution of the parasitoid species. Total nestedness is given, when species diversity in all smaller habitats are only subsamples of the diversity in the largest habitat, and thus all species occurring in the small habitats can also be found in the largest habitat (Wright & Reeves, 1992; Atmar & Patterson, 1993). A strongly nested pattern may be caused by negative correlations between immigration and extinction (Brown & Kodric-Brown, 1977). A low nested distribution pattern, as we found for the parasitoids, leads to a high diversity between habitats (called ß-diversity, Whittaker, 1972), caused by the occurrence of "new" species in the species-pools of small habitats lacking in the more diverse large habitats.

Three parasitoid species of the vetch insects occurred only in the smallest meadow (*Pristomerus vulnerator*, *Glabrobracon* sp.2, *Trichomalus repandus*). This may be due to broader host ranges, since at least two of the species (*P. vulnerator* and *Glabrobracon* sp.2 both parasitizing the moth *Cydia nigricana* in the pods) are polyphagous on different Microlepidoptera species. So, availability of alternative hosts in adjoining habitats may have influenced parasitoid persistence in isolated habitat islands.

In contrast to the parasitoids, the phytophagous insects on *T. pratense* and *V. sepium* showed a totally nested distribution pattern since all species in the small meadows were also present in large meadows. Thus, cumulative species-area curves were identical for both cumulation methods shown above.

In consequence, our results on parasitoid communities underlines the importance of "several small" habitats for species diversity. But this is only on the basis of presence/absence data. Species abundance or ecosystem functions like parasitism are not considered. However, our results on the parasitism of phytophagous insects (Fig. 2a, 2c) show that the "several small" alternative would lead to a reduction of parasitism pressure on the herbivores.

4.7 Summary

Our investigations in plant-insect communities showed that fragmentation of habitats negatively affects species diversity, species abundance, and species interactions of endophagous herbivores and their parasitoids. The two major features of fragmentation, decrease in area and increase in isolation, led to nearly identical effects on both investigated plant-insect systems (endophagous insects on *Trifolium pratense* and *Vicia sepium*).

Species diversity was dramatically reduced by both area-loss and increasing habitat isolation. Whereas the decrease in species number in the meadows are likely due to local species extinction, lower species diversity on the experimentally created isolated plant plots are probably due to colonization failures.

Species-area relationships on the meadows were much stronger for parasitoids than for herbivores, due to a less nested distribution of parasitoids but not of herbivores across the different meadows. Multiple regression analysis showed that the steep decline in parasitoid diversity with habitat fragmentation was not related to host abundance or host species diversity.

Herbivore diversity was less affected by isolation than parasitoid diversity, and habitat isolation by 100 meters or more led to a steep decrease in species richness of parasitoids, but not in herbivores.

The absence of local species in both the old meadows and the isolated plant plots was found to be closely correlated with species abundance and population variability. This primarily affected the parasitoids since they had lower abundances and higher population variabilities than their hosts.

Parasitism pressure on the herbivorous insects in both systems were dramatically affected by both area reduction and increase of habitat isolation. Hence herbivores on small and isolated habitats were greatly released from parasitism.

Cumulative species-area curves of parasitoids supported the "several small" strategy to increase parasitoid diversity. But probability of local species extinctions was higher in small and isolated habitats since population density of parasitoids was reduced and population variability increased. Thus "single large" habitats are essential for stabilizing species diversity on a metapopulation level, since in small and isolated habitats the time between population crashes may be shorter than the recovery time, especially for specialized parasitoid species. Thus, in the agricultural landscape, enhancement of both abundance and species diversity of biological control agents, necessitates a landscape design with both "some large and several small" habitats (SLASS). Both strategies have merits.

Acknowledgements

Taxanomic advice and help with species identification came from B. Büche (Coleoptera), M. Čapek (Braconidae), K. Horstmann (Ichneumonidae), H. Meyer (Cecidomyiidae), W. Sauter (Tortricidae), and S. Vidal (Chalcididae). Comments by John Banks improved the manuscript.

4.8 References

Andrén, H., 1992. Corvid density and nest predation in relation to forest fragments: a landscape perspective. *Ecology* **73**: 794-804.

Andrén, H., 1994. Effects of habitat fragmentation on birds and mammals in landscapes with different proportions of suitable habitat: a review. *Oikos* **71**: 355-366.

Andrén, H., 1996. Population responses to habitat fragmentation: statistical power and the random sample hypothesis. *Oikos* **76**: 235-242.

Atmar, W. & Patterson, B.D., 1993. The measure of order and disorder in the distribution of species in fragmented habitat. *Oecologia* **96**: 373-382.

Bascompte, J. & Solé, R.V., 1996. Habitat fragmentation and extinction thresholds in spatially explicit models. *J. Anim. Ecol.* **65**: 465-473.

Baur, B. & Erhardt, A., 1995. Habitat fragmentation and habitat alterations: principal threats to most animal and plant species. *Gaia* **4**: 221-226.

Brown, J.H., 1971. Species richness of boreal mammals living on the montane islands of the great basin. *Am. Nat.* **105**: 467-478.

Brown, J.H., 1986. Two decades of interaction between the MacArthur-Wilson model and the complexities of mammalian distributions. *Biol. J. Linn. Soc.* **28**: 231-251.

Brown, J.H. & Kodric-Brown, A., 1977. Turnover rates in insular biogeography: effects of immigration on extinction. *Ecology* **58**: 445-449.

Burgess, R.L. & Sharpe, D.M. (Eds.), 1981. *Forest island dynamics in man-dominated landscapes*. Springer, New York.

Burkey, T.V., 1989. Extinction in nature reserves: the effect of fragmentation and the importance of migration between reserve fragments. *Oikos* **55**: 75-81.

Connor, E.F. & McCoy, E.D., 1979. The statistics and biology of the species-area relationship. *Am. Nat.* **113**: 791-833.

Davis, B.N.K., 1975. The colonization of isolated patches of nettles (*Urtica dioica* L.) by insects. *J. Appl. Ecol.* **12**: 1-14.

Davis, B.N.K. & Jones, P.E., 1986. Insects on isolated colonies of common rock-rose Helianthemum chamaecistus. *Ecol. Entomol.* **11**: 267-281.

Diamond, J.M. & May, R.M., 1981. Island biogeography and the design of natural reserves. In: May, R.M. (Ed.), *Theoretical Ecology*. Blackwell Scientific Publ., Oxford.

Didham, R.K., Hammond, P.M., Lawton, J.H., Eggleton, P. & Stork, N.E., 1998. Beetle species responses to tropical forest fragmentation. *Ecological Monographs* **68**:295-323.

Dieckmann, L., 1977. Beiträge zur Insektenfauna der DDR: Coleoptera - Curculionidae (Apioninae). *Beitr. Entomol.* **27**: 7-143.

Ferrari, J., Kruess, A. & Tscharntke, T., 1997. Auswirkungen der Fragmentierung von Wildrosen auf deren Insektenlebensgemeinschaften. *Mitt. Dtsch. Ges. Allg. Angew. Ent.* **11**: 87-90.

Freude, H., Harde, K.W. & Lohse, G.A., 1981. Die Käfer Mitteleuropas Bd. 10: Bruchidae, Anthribidae, Scolytidae, Platypodidae, Curculionidae. Goecke & Evers, Krefeld.

Gilpin, M.E., 1987. Spatial structure and population vulnerability. In: Soulé, M.E. (Ed.), *Viable populations for conservation*. Cambridge Univ. Press, pp. 125-139.

Gilpin, M.E.& Hanski, I., (Eds.), 1991. *Metapopulation dynamics: Empirical and theoretical investigations*. London, Academic Press.

Gilpin, M.E. & Soulé, M.E., 1986. Minimum viable populations: the processes of the species extinctions. In: Soule, M.E. (Ed.), *Conservation Biology: The science of scarcity and diviersity*. Sinauer, Sunderland, MA., pp. 13-34.

Gustavson, E.J. & Parker, G.R., 1992. Relationships between landcover proportion and indices of landscape spatial pattern. *Landscape Ecol.* **7**: 101-110.

Haila, Y., 1983. Land birds on northern islands: a sampling metaphor for insular colonization. *Oikos* **41**: 334-351.

Haila, Y. & Järvinen, O., 1983. Land bird communities on a Finnish island: species impoverishment and abundance patterns. *Oikos* **41**: 255-273.

Hanski, I., Kuusaari, M. & Nieminen, M., 1994. Metapopulation structure and migration in the butterfly *Melitaea cinxia*. *Ecology* **75**: 747-762.

Harris, L.D., 1984. *The fragmented forest: island biogeography theory and preservations of biotic diversity*. Chicago, University Chicago Press.

Have, A., 1993. Effects of area and patchiness on species richness: an experimental archpelago of ciliate microcosmos. *Oikos* **66**: 493-500.

Hawkins, B.A. & Gross, P., 1992. Species richness and population limitation in insect parasitoid-host systems. *Am. Nat.* **139**: 417-423.

Hawkins, B.A., Thomas, M.B. & Hochberg, M.E., 1993. Refuge theory and biological control. *Science* **292**: 1429-1437.

Hölzinger, J., 1987. *Die Vögel Baden-Württembergs. I. Gefährdung und Schutz*. Ulmer, Stuttgart.

Hurka, H., 1984. Influence of population parameters on the genetic structure of *Capsella* populations. In: Wöhrmann, K. & Loeschke, V. (Eds.), *Population Biology and Evolution*. Springer, Berlin.

Johnson, M.P. & Simberloff, D.S., 1974. Environmental determinants of island species numbers in the British Isles. *J. Biogeogr.* **1**: 149-154.

Kareiva, P., 1987. Habitat-fragmentation and the stability of predator-prey interactions. *Nature* **326**: 28-290.

Kareiva, P., 1990. Population dynamics in spatially complex environments: theory and data. *Philosophical Transactions of the Royal Society of London Series B* **330**: 175-190.

Kareiva, P. & Wennergren, U., 1995. Connecting landscape patterns to ecosystems and population processes. *Nature* **373**: 299-301.

Kimberly, A. W. & Christ, T. O., 1995. Critical thresholds in species' responses to landscape structure. *Ecology* **76 (8)**: 2446-2459.

Kruess, A., 1996. Folgen der Lebensraum-Fragmentierung für Pflanze-Herbivor-Parasitoid-Gesellschaften: Artendiversität und Interaktionen. *Agrarökologie* **18**: 1-134.

Kruess, A. & Tscharntke, T., 1994. Habitat fragmentation, species loss, and biological control. *Science* **264**: 1581-1584.

Kruess, A. & Tscharntke, T., 1999. Species richness and parasitism in a fragmented landscape: experiments and field studies on *Vicia sepium*. *Oecologia* (in press).

Lawton, J.H., 1995. Population dynamic principles. In: Lawton, J.H. & May, R.M. (Eds.), *Extinction rates*, Oxford, Oxford University Press, pp. 147-163.

Loman, J. & von Schantz, T., 1991. Birds in a farmland - more species in small than in large habitat islands. *Cons. Biol.* **5**: 176-188.

MacArthur, R.H. & Wilson, E.O., 1967. *The theory of island biogeography*. Princeton University Press, Princeton.

MacGarvin, M., 1982. Species-area relationships of insects on host plants: herbivores on rosebay willowherb. *J. Anim. Ecol.* **51**: 207-223.

Margules, C.R., Milkovits, G.A. & Smith, G.T., 1994. Contrasting effects of habitat fragmentation on the scorpion *Cercophonius squama* and an amphipod. *Ecology* **75**: 2033-2042.

Martin, J.L. & Lepart, J., 1989. Impoverishment in the bird community of a Finnish archipelago: the role of island size, isolation and vegetation structure. *J. Biogeogr.* **16**: 159-172.

Noss, R.F., 1991. Landscape connectivity: Different functions at different scales. In: Hudson, W.E. (Ed.), *Landscape linkages and biodiversity*. Washington, D.C., Island Press.

Pimm, S.L., 1991. *The balance of nature? Ecological issues in the conservation of species and communities*. University of Chicago Press, Chicago.

Quinn, J.F. & Harrison, S.P., 1988. Effects of habitat fragmentation and isolation on species richness: evidence from biogeographic patterns. *Oecologia* **75**: 132-140.

Rigby, C. & Lawton, J.H., 1981. Species-area relationships of arthropods on host plants: herbivores on bracken. *J. Biogeogr.* **8**: 125-133.

Rosenzweig, M.L., 1995. *Species diversity in space and time*. University Press, Cambridge.

Saunders, D.A., Hobbs, R.J. & Arnold, G.W., 1993. The Kellerberrin project on fragmented landscapes:

a review of current information. *Biol. Conserv.* **64**: 185-192.

Schaal, B.A. & Leverich, W.J., 1984. Age-specific fitness components in plants: genotype and phenotype. In: Wöhrmann, K. & Loeschke, V. (Eds.), *Population Biology and Evolution.* Springer Verlag, Berlin.

Shaffer, M.L., 1981. Minimum population sizes for species conservation. *Bioscience* **31**: 131-134.

Simberloff, D., 1988. The contribution of population and community biology to conservation science. *Ann. Rev. Ecol. Syst.* **19**: 473-511.

Soulé, M.E., 1987. *Viable populations for conservation.* Cambridge University Press, Cambridge.

Spiller, D.A. & Schoener, T.W., 1988. An experimental study of the effects of lizards on web-spider communities. *Ecol. Monogr.* **58**: 57-77.

Taylor, A.D., 1990. Metapopulations, dispersal, and predator-prey dynamics: an overview. *Ecology* **71**: 429-433.

Thomas, C.D. & Harrison, S., 1992. Spatial dynamics of a patchily distributed butterfly species. *Journal of Animal Ecology* **61**: 437-446.

Thomas, C.D., Thomas, J.A. & Warren, M.S., 1992. Distribution of occupied and vacant butterfly habitats in fragmented landscapes. *Oecologia* **92**: 563-567.

Thornton, I.W.B., Zann, R.A. & van Balen, S., 1993. Colonization of Rakata (Krakatau Is.) by non-migrant land birds from 1883 to 1992 and implications for the value of island equilibrium theory. *J. Biogeogr.* **20**: 441-452.

Tjernberg, M., Johnsson, K. & Nilsson, S.G., 1993. Density variation and breeding success of the black woodpecker *Dryocopus martius* in relation to forest fragments. *Ornis Fennica* **70**: 155-162.

Tscharntke, T. & Kruess, A., 1999. Habitat fragmentation and biological control. In: Hawkins, B.A. & Cornell, H.V. (Eds.), *Theoretical approaches to biological control*; Cambridge University Press, Cambridge, pp. 190-205.

Turner, M.G., 1989. Landscape ecology: The effect of pattern and process. *Ann. Rev. Ecol. Syst.* **20**: 171-197.

Villard, M.A. & Taylor, P.D., 1994. Tolerance to habitat fragmentation influences the colonization of new habitat by forest birds. *Oecologia* **98**: 393-401.

Ward, L.K. & Lakhani, K.H., 1977. The conservation of Juniper: The fauna of food-plant island sites in Southern England. *J. Appl. Ecol.* **14**: 121-135.

Whittaker, R.H., 1972. Evolution and the measurement of species diversity. *Taxon* **21**: 213-251.

Wilcox, B.A, 1980. Insular ecology and conservation. In: Soulé, M.E. & Wilcox, B.A. (Eds.), *Conservation biology: an evolutionary-ecological perspective.* Sinauer, Sunderland, MA., pp. 95-117.

Wilcox, B.A. & Murphy, D. D., 1985. Conservation strategy: the effects of fragmentation on extinction. *Am. Nat.* **125**: 879-887.

Wilcove, D.S., McLellan, C.H. & Dobson, A.P., 1986. Habitat fragmentation in the temperate zone. In: Soulé, M.E. (Ed.), *Conservation Biology. The science of scarcity and diversity.* Sinauer, Sunderland, MA., pp. 237-256.

Williams, C.B., 1964. *Patterns in the balance of nature and related problems in quantitative ecology.* New York, Academic Press.

Wilson, E.O. & Willis, E.O., 1975. Applied biogeography. In: Cody, M.L. & Diamond, J.M. (Eds.), *Ecology and evolution of communities*; Cambridge, Mass., Harvard University Press, pp. 522-534.

With, K.A. & Christ, T.O., 1995. Critical threshold in species responses to landscape structure. *Ecology* **76**: 2446-2459.

Wright, D.H. & Reeves, J.H., 1992. On the meaning and measurement of nestedness of species assemblages. *Oecologia* **92**: 416-428.

Zabel, J. & Tscharntke, T., 1998. Does fragmentation of *Urtica* habitats affect phytophagous and predatory insects differentially? *Oecologia* **116**: 419-425

CHAPTER 5

THE IMPACT OF CORRIDORS ON ARTHROPOD POPULATIONS WITHIN SIMULATED AGROLANDSCAPES

GARY W. BARRETT
Institute of Ecology, University of Georgia, Athens, GA USA

5.1 Introduction

Landscape ecology is a relatively new integrative field of study that weds ecological theory with practical application (Barrett & Bohlen, 1991). Specifically, landscape ecology considers the development and dynamics of spatial heterogeneity, spatial and temporal interactions and exchanges across heterogeneous landscapes, influences of spatial heterogeneity on biotic and abiotic processes, and the management of spatial heterogeneity (Risser *et al.*, 1984). A primary focus of landscape ecology is how a heterogeneous geographic area (*e.g.*, a large region of conventional row-crop agriculture) can best be managed to simultaneously maximize both ecological and societal benefits (*e.g.*, crop productivity, public recreation, biotic diversity, insect pest control, and nutrient recycling, among others). Traditionally, a single field (agroecosystem) approach was employed to address questions and to solve problems related to concepts or topics such as Integrative pest management, restoring biotic diversity, or improving crop yield.

Further, scientists traditionally employed only a single methodology - the scientific method - to address these questions and to solve these problems. Ecologists and resource managers have often failed to recognize that several research approaches (*e.g.*, cost-benefit analysis, net energy, cybernetics, and problem-solving algorithms, among others) are available for resource-management, impact assessment, and hypothesis-testing progress (Barrett, 1985).

Just as we learned during the past two decades that biotic diversity cannot be conserved by a single species approach (Salwasser, 1991), hopefully we will learn as we get ready to enter the 21st century that we cannot sustain agricultural productivity by a single field (agroecosystem) approach. Rather, an agrolandscape approach is needed in which landscape elements (*e.g.*, patches and corridors) are patterned to optimize for a set of objectives related to insect pest control, nutrient restoration, habitat fragmentation, trophic and biotic diversity, primary productivity (natural and subsidized), and connectivity (Barrett, 1992). Thus, a new field of study - agrolandscape ecology - must continue to evolve if ecologists and resource managers are to manage agriculture in a sustainable manner for future generations. It is imperative that ecologists find ecological solutions to challenges such as how best to reduce insect pest damage, to promulgate management strategies such as Integrative Pest Management (IPM), and to implement research findings at greater temporal/spatial scales such as the Sustainable Biosphere Initiative (SBI) (Lubchenco,

B. Ekbom, M. Irwin and Y. Robert (eds.), Interchanges of Insects, 71-84

et al., 1991; Barrett, 1994). This chapter will focus on the impact of mainly one component of this landscape matrix, namely landscape corridors, and how this component affects arthropod population densities and patterns of movement within simulated, experimental agrolandscapes. I suggest that an experimental approach is necessary to evaluate and to more fully understand the role of corridors on arthropod populations. The simulated-landscape research design must be replicated to avoid the problems of pseudoreplication (Hurlbert, 1984). A new integrative perspective is needed to address questions and to solve problems at the agrolandscape scale. Before turning to actual research designs and field results, perhaps a few comments are in order concerning this perspective.

5.2 An Integrative Agrolandscape Perspective

To better understand arthropod movement patterns, one needs to better understand the architecture and geometry of the landscape mosaic (*e.g.*, the role of patch size and landscape corridors in the landscape mosaic), and how landscape processes such as arthropod dispersal behavior are affected by landscape structure and habitat fragmentation (Barrett & Bohlen, 1991). Increasingly, there is evidence that a holistic (top-down) approach should be viewed as complementary to a reductionist (bottom-up) approach (*e.g.*, Carpenter & Kitchell, 1988, 1993) if ecologists are to advance our knowledge concerning such phenomena as arthropod dispersal behavior and patterns of movement. This knowledge will also help to insure that insect pest management strategies are integrated in an ecologically-efficient and cost-effective manner. This new perspective requires that theory and application be coupled in a holistic research and problem-solving management approach (Barrett, 1985).

An array of research concepts (*e.g.*, hierarchy and landscape theory) and monitoring technologies (*e.g.*, GIS and systems analysis) should be formulated according to the questions being addressed, the temporal/spatial scale to be evaluated, and the resource management goals to be implemented. This approach and perspective must also include an integration of landscape theory with resource management goals, an understanding of historical and predicted disturbance regimes, and a recognition of socio-economic constraints (Barrett & Bohlen, 1991).

Hypotheses to be tested or problems to be solved are frequently evaluated at the wrong spatial scale (*i.e.*, evaluated at the population or community levels rather than the ecosystem or landscape levels). Further, problems are all too often assessed at the wrong temporal scale (*i.e.*, assessed in terms of short-term budgetary constraints rather than in terms of long-term sustainable benefits). Problems frequently arise, for example, when perturbations (*e.g.*, a pesticide application) are tested at one level (the population level) and then applied without sufficient study at another level (the ecosystem or landscape level). These differences are also dependent on the trophic levels affected (Barrett, 1968), on biotic diversity and plant life histories (Carson & Barrett, 1988), and on how transcending processes differ within and across levels of organization (Barrett *et al.*, 1997).

A better understanding of arthropod movement patterns, therefore, requires that a holistic (landscape) approach be integrated with a reductionist (population dynamic) approach. Studies designed and hypotheses tested based on a simulated and replicated landscape perspective provide an effective means to address questions at greater temporal/spatial scales. Further, there is frequent need to design studies aimed at restoring landscape elements (Baldwin *et al.*, 1994; Barrett, 1994). The emerging field of restoration ecology focuses on restoration needs and case study approaches (*e.g.*, Peles *et al.*, 1996). An objective of understanding arthropod patterns of movement will increasingly require cross- and transdisciplinary approaches focused on large landscape units (*e.g.*, a watershed) that encompass all landscape elements (patches, corridors, matrix, and human built structures). This perspective and research agenda should also include how to restore and how to structure these watershed or landscape units based on long-term sustainablity (*i.e.*, using our understanding of mature sustainable natural ecosystems or landscapes as model systems for reference) if we are to integrate humankind within the landscape concept (Odum, 1969; Barrett, 1989, 1992).

This chapter will mainly focus on how we can manipulate and/or structure a landscape element - corridors - when investigating or designing a more sustainable agricultural landscape, and how these corridors affect insect population dynamics, rates of dispersal, and patterns of movement. When designing such studies it is imperative to note that corridors may manifest both ecological/economic benefits (increased biotic diversity and reduced soil erosion), as well as potential harmful effects (transmission of disease or decreased crop yield). Therefore, concepts such as cost-benefit analysis, net energy, and net profit must be viewed in a transdisciplinary and holistic manner in order to arrive at landscape designs and to implement restoration practices that will benefit society on a sustainable basis.

5.3 Landscape Corridors - Types and their Relationship to Insect Movement

Natural corridors (*e.g.*, stream corridors) have always been an important component of the landscape. Human-established corridors (*e.g.*, fencerows, hedgerows, and roadside vegetation) have also been a major element of the fragmented landscape for decades. Planted corridors, for example, were established in the Great Plains of the United States in the 1930's to reduce wind erosion and provide wood for fuel (Shelterbelt Project, 1934).

There presently exists five basic types of corridors based on their origin: disturbance corridors, planted corridors, environmental resource corridors, regenerated corridors, and remnant corridors.

Disturbance corridors (*e.g.*, power line cuts) disrupt the natural, more homogenous landscape. Disturbance corridors act as barriers to movement of some species, but provide dispersal routes for several species of insects, birds, and mammals. More recently, management practices of disturbance corridors have encompassed the heterogeneity of corridor vegetation and, consequently, provide habitat for nesting birds, food resources for game species, and niches for insects and small mammals.

Planted corridors (*e.g.*, shelterbelts, see above) are common in agriculture landscapes to prevent soil erosion, to provide habitat for wildlife, and to enhance biotic diversity.

Environmental resource corridors (*e.g.*, a riparian forest along a stream) are important to intercept nutrients and sediments from agricultural run-off that would otherwise end up in streams causing cultural eutrophication; the riparian zone also reduces extreme fluctuations in stream beds.

Regenerated corridors (*e.g.*, strips of vegetation along roadsides that regenerate from a previously disturbed area during secondary succession) are common in the midwestern United States and in most European countries. Regenerated corridors provide refuge for plant and animal populations, and also serve as important links that allow a diversity of animal movement and seed dispersal between larger habitat patches.

Remnant corridors (*e.g.*, a strip of native vegetation) are probably the most important (but not the most common) type of corridor for conserving biotic diversity, closing nutrient cycles, and maximizing landscape stability. Remnant corridors depict how "mature" corridors likely have functioned for centuries (*i.e.*, serve as a control model), in contrast to corridors which have either been established or severely impacted by humankind.

Each of these five types of corridors provide niches for insects and their predators, impact insect movement, and affect rates of dispersal behavior. Although several investigations have focused on long-term insect patterns of movement (see review by Stinner *et al.*, 1983 for details), few studies have been designed to address the role of corridor type on patterns of movement by insects based on a replicated, simulated-landscape research design. The following experimental research designs and recommendations are intended to illustrate how ecologists and resource managers might address questions and test hypotheses at the agrolandscape level of organization.

5.4 Experimental Research Designs

Landscape corridors have recently been recognized as significant elements in the landscape mosaic (Forman & Godron, 1981; Barrett & Bohlen, 1991). For example, both natural and human-built corridors have been experimentally used to investigate the effect of corridors on small mammal population dynamics and patterns of movement (Lorenz & Barrett, 1990; La Polla & Barrett, 1993; Williams *et al.*., 1994). Barrett *et al.* (1995), Diffendorfer *et al.* (1995), and Barrett & Peles (1999) present an overview regarding the use of experimental landscapes in mammalian ecology.

Likewise, experimental landscape corridors have been used to investigate the role of corridors on arthropod population dynamics and patterns of movement (Forman & Baundry, 1984; Kemp & Barrett, 1989). Figure 1A illustrates the replicated, large-scale research design employed by Kemp & Barrett (1989) and Rodenhouse *et al.* (1992). Kemp & Barrett (1989) determined that uncultivated, grassy corridors within soybean agroecosystems (Fig. 1B) reduced the densities and affected the distribution of adult potato leafhoppers (*Empoasca fabae*), and increased the role of infestation of the green cloverworm (*Plathypena scrabra*) by the fungal pathogen (*Nomuraea rileyi*). Uncultivated successional corridors (*i.e.*, corridors dominated by early successional old-field plant species), however, failed to "funnel" predaceous arthropods into the soybean crop, although predators were more abundant in these uncultivated successional corridors. Experimental grassy corridors also demonstrated that grasses adjacent to or within the soybean crop conferred an "associational resistance" (Altieri, 1977; Schoonhoven *et al.*, 1981) to movement by certain insect species.

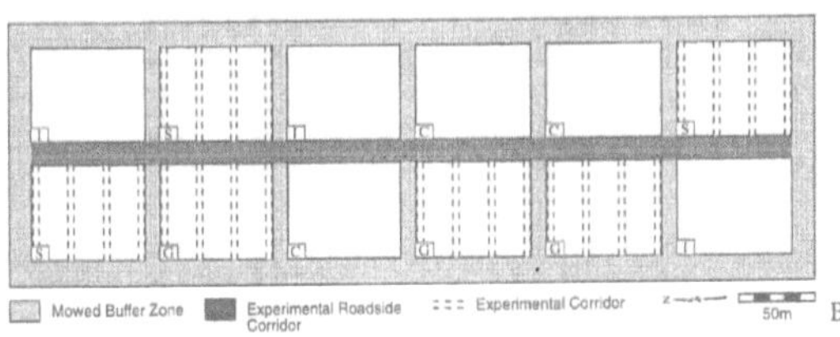

Figure 1. Aerial photograph of the research site (A), including a diagram depicting the experimental research design (B), and a photo of a grassy corridor (C). Photographs taken during August, 1985.

Rodenhouse *et al.* (1992) further investigated the effects of uncultivated corridors on arthropod abundance and movement patterns in soybean agroecosystems. They found that corridors suppressed populations of leaf and stem-sucking pests, particularly *Empoasca fabae* but not defoliators which were often more abundant in soybean plots with corridors than in control plots. Predaceous arthropods were more abundant in successional than grassy corridors, although soybean yields per meter of crop row did not differ significantly between controls and other treatments. They noted, however, because of the multiple benefits of landscape corridors, as noted earlier in this chapter, that uncultivated corridors should be established within croplands for integrated pest management purposes. These earlier findings also raised the question regarding the movement of insects within and across soybean plots which were strip intercropped with other types of agricultural crops.

Bohlen & Barrett (1990) released and recaptured marked Japanese beetles, *Popillia japonica*, in experimental plots to test the effects of contrasting types of strip-cropped soybean agroecosystems on beetle dispersal behavior. Experimental treatments were soybean monoculture, soybean strip-cropped with dwarf soybean, and soybean strip-cropped with tall soybean. Beetles remained longer in the center of the strip-cropped bicultures than in the monoculture indicating that the strips (corridors) of sorghum inhibited their movement. Rates of dispersal, however, were similar in dwarf and tall soybean treatments despite differences in plant height. This finding further strengthened the *associational resistance* hypothesis (see above) as a mechanism to regulate insect populations at the landscape scale. There was also evidence that tall soybean oriented beetles to move parallel to intercropped strips based on the number of beetles captured at the end of each experimental plot. These findings suggested that strip intercropping can affect the movement and dispersal behavior of polyphagous herbaceous insects and provide benefits for pest management in large scale agroecosystems.

Holmes & Barrett (1997) designed a large scale agroecosystem study to further investigate if strips of short soybean would indeed significantly affect the abundance, rates of dispersal, and patterns of movement of Japanese beetles across experimental plots of soybeans. Marked and unmarked populations of Japanese beetles were monitored using trap and direct observation census methods in a replicated field research design. This replicated research design (Fig. 2A) was intended to simulate a large-scale agrolandscape level of investigation. Pheromone beetle traps were situated in the center of both monoculture soybean plots (Fig. 2B) and the center of soybean plots strip intercropped with sorghum (Fig. 2C).

They found significantly lower densities of Japanese beetles in the intercropped treatment compared to the monoculture treatment. Rates of dispersal were also significantly decreased in the intercropped treatment, suggesting that strip intercropping should be a component of integrated pest management in the control of generalist, herbivore insect species.

Overall, 13.5% of all marked beetles were recaptured in the Holmes & Barrett (1997) study, with only 1.7 % recaptured in a distance plot. The farthest a recaptured beetle emigrated from the site of release was 400 m. They did recommend, however, that future experimental studies of this species be established at even greater spatial scales. Greater temporal and spatial scales should be factored into research designs when investigating insect taxonomic groups such as Lepidoptera (Ehrlich & Murphy, 1987;

Figure 2. Aerial photograph of the research site and design (A), including photographs of a soybean monoculture (B), and a sorghum–soybean strip intercropped (C) agroecosystem. Photographs taken during July, 1993.

New, 1991), Hymenoptera (Pavuk & Barrett, 1993), Hemiptera (Rodenhouse *et al.*, 1992), and Coleoptera (Kemp & Barrett, 1989; Bohlen & Barrett, 1990). Well designed experiments need to cover a range of spatial and temporal scales and need to be sufficiently replicated to provide statistically robust results (Nichols & Margules, 1991; Fahrig & Paloheimo, 1988; Pulliam *et al.*, 1992).

5.5 The Twenty-First Century

The twentieth century has witnessed a greater understanding of arthropod patterns of abundance at the field (*i.e.*, at the agroecosystem) level, but has not achieved a clear understanding of arthropod patterns of movement or dispersal behavior at the agrolandscape level (Barrett, 1992). Likewise, there have been numerous break-throughs in areas of research such as plant-arthropod (co-evolutionary) interactions as related to biodiversity (see review by Reaka-Kudla *et al.*, 1997 for details), in our understanding of and need to implement alternative agriculture (National Research Council, 1989) at greater spatial scales, and in the need to conserve biotic diversity at the cellular, species, and habitat levels of organization (Barrett *et al.*, 1997). However, we urgently need to increase our understanding of topics such as the role of corridors on arthropod dispersal behavior at the landscape level, the optimum geometry and fragmentation of the landscape to simultaneously balance crop yield with biotic diversity, and how best to integrate humankind (*i.e.*, socio-economic factors) with the agroecosystem and agrolandscape concepts.

Listed below are five nonprioritized recommendations that are intended to integrate educational, research, service, policy, and management strategies as we as a society prepare to enter the 21st century. Examples and investigations involving the role of landscape corridors on arthropod population dynamics and dispersal behavior will be used to illustrate why we need to consider, and hopefully implement, these recommendations. These recommendations should also increase our understanding regarding the integration of humankind with the agrolandscape concept.

Scientists, resource managers, and policy makers need to address questions regarding insect movement and integrated (sustainable) pest management at greater temporal/spatial scales. The importance of long-term ecological research at the ecosystem level has long been recognized (Callahan, 1984; Likens, 1989). Indeed, a network of Long-term Ecological Research (LTER) sites funded by the National Science Foundation has been in operation for over a decade. It is only more recently, however, that scientists have recognized the need to scale such investigations to the landscape and global levels (Lubchenco *et al.*, 1991; Allen & Hoekstra, 1992). A landscape approach provides a holistic perspective which helps to insure that cost-effective resource management and insect pest control occurs in a coordinated manner and that concepts such as sustainablity, carrying capacity, connectivity, net energy, and minimum critical scale are encompassed in the decision making process (Barrett & Bohlen, 1991).

For example, the application of an insecticide at the agroecosystem (crop field) level may suffice as a short-term solution to control a particular insect pest such as the Mexican bean beetle, *Epilachna varivestis*. However, a landscape approach, including the mosaic of cropland and forest patches, is needed to address insect control of this pest species since Mexican bean beetles are known to overwinter in forest or woodland patches in the total landscape mosaic (Kogan & Kuhlman, 1982). Thus there is increased recognition that an integrated agroecosystem management approach (El Titi & Landes, 1990), coupled with an agrolandscape (Barrett, 1992; Peles & Barrett, 1994) approach, be employed in the control of and research focusing on integrated pest management strategies.

Scientists, resource managers, and policy makers need to address questions regarding insect movement and integrated (sustainable) pest management in a transdisciplinary manner. While disciplinary (specialization in isolation) and multidisciplinary (multiple approaches without cooperation) approaches will continue to advance science at the cellular, organismic, and population levels of organization, interdisciplinary (coordination by a higher level concept) and transdisciplinary (multi-level coordination of an entire system) approaches are urgently needed to address questions at the ecosystem, landscape, and ecosphere levels of investigation (see Jantsch, 1972 and Johnson, 1977 for detailed information regarding these approaches).

For example, landscape corridors and strip intercropping have been shown to be effective in the control of select insect pests such as the potato leafhopper (*Empoasca fabae*) and the Japanese beetle (*Popillia japonica*) (Kemp & Barrett, 1989; Rodenhouse *et al.*, 1992; Holmes & Barrett, 1997). To effectively implement research and insect pest control programs at the agrolandscape level will require transdisciplinary cooperation (managers, scientists, educators, policy makers, and the citizens) at greater temporal and spatial scales.

Scientists, resource managers, and policy makers need to address questions regarding insect movement and integrated (sustainable) pest management based on transcending processes across levels of organization. Ecologists have long recognized the importance of the levels-of-organization concept (see Rowe, 1961 and MacMahon *et al.*, 1978 for historical reviews) concerning the integration and organization of information in a hierarchical (Allen & Starr, 1982; O'Neill *et al.*, 1986) manner. A hierarchy is defined as an arrangement into graded series of components arranged from the largest to the smallest, but the order could be reversed if one wishes to start with the lowest level of resolution (Odum, 1997). The levels of organization concept allows one to investigate the natural world in terms of increasing complexity from the molecular or cellular levels through the ecosystem, landscape, and ecosphere levels. There is also urgent need to encompass humankind into the levels of organization concept (Barrett, 1985; Odum, 1997). Recently Barrett *et al.* (1997) stressed the need to teach and address problems regarding those principles, natural laws, mechanisms, and processes that transcend all levels of organization. These include, among others, energetics, evolution, regulation, and diversity.

For example, in order to investigate the role of diversity (genetic, species, or habitat) regarding the impact of landscape corridors on insect dispersal behavior or integrated pest management strategies, one would be wise to integrate information and address this question from the molecular or cellular through the ecosystem, landscape, and ecosphere levels. The role of landscape corridors in alternative agriculture, for example, will require the integration of information ranging from the effects of microhabitat conditions predaceous insects to the use of Geographic Informational Systems (GIS) in monitoring patterns of movement at the landscape level.

Scientists, resource managers, and policy makers need to address questions regarding insect movement and integrated (sustainable) pest management based on the concept of optimizing habitat fragmentation. During the past several decades conventional agricultural management mainly focused on the single field crop or at the agroecosystem level. This approach encouraged the use of high technology and increased subsidies (fossil fuels, fertilizers, and pesticides) aimed primarily at increasing crop yield (NRC, 1989). More recently, however, fields of study such as landscape ecology, conservation biology, and restoration ecology have provided vital information and recent insights resulting in new approaches to sustainable agriculture at the agrolandscape level (Barrett 1992; Barrett & Peles, 1994). Landscape ecology, for example, weds ecological theory with practical application at greater scales (*i.e.*, at the regional or watershed scale). At this scale, questions resulting from investigating the impact of increased land use patterns on habitat fragmentation within the human populated landscape, such as in the agricultural Midwest in the United States, have received increased attention (*e.g.*, Barrett *et al.*,1990).

Theoretically, there exists an optimum degree of fragmentation in an agrolandscape to balance (and connect) natural landscape elements, such as habitat patches and corridors, with those areas devoted to crop productivity - especially crop productivity based on the concept of sustainable agriculture. If societies are to establish a truly sustainable type of agriculture, and conserve or restore biodiversity on a long-term basis (Noss, 1983, 1991), then the optimum design of the landscape mosaic deserves increased attention. This challenge involves the integration of ecosystem and landscape science with society (Pastor, 1995; Seastedt, 1996). I predict the 21st century will increasingly witness the application of ecological principles and concepts to questions involving the most cost-effective and ecologically-safe design and management of landscapes at this scale. Questions regarding the role of corridors on insect patterns of movement will simultaneously be elucidated as part of this sustainable, integrative pest management strategy.

Lastly, there is urgent need for scientists, resource managers, and policy makers to address questions regarding insect movement and integrated (sustainable) pest management based on public ecological literacy. As we address such questions as the role of landscape corridors on insect patterns of dispersal, including how these data can be encompassed into the concept of sustainable pest control and sustainable development at greater scales, it is imperative that society command the knowledge concerning how ecological systems function (in which they are a component part) in order to make decisions that will sustain resources and conserve biodiversity for generations to come. Sustainablity is, perhaps, best defined as "maintenance of natural capital" (Goodland, 1995). A society that participates in and makes decisions based on the concept of sustainablity will likely be able to conserve this natural capital (resources and services) and even enhance the human designed landscape in which they function, especially as we design with, rather than against, natural ecological processes. It is vital that society understand processes and concepts such as nutrient recycling, carrying capacity, net energy, and optimum habitat fragmentation and connectivity, among others, to implement this sustainable approach to resource management. A society educated in this manner will likely conduct such functions as recycling, energy conservation, and pest management based on an educational incentive rather than on a regulatory mandate (Barrett, 1989).

Therefore, the topic of this book - Interchanges of Insects Between Agricultural and Surrounding Landscapes - is most appropriate in order to outline strategies regarding why researchers, insect pest control agents, industry, policy makers, educators, and the private land owner must work together in a transdisciplinary context if we are to more fully understand and more effectively manage our biotic and abiotic resources in a sustainable manner. Such a sustainable approach should provide an agenda for and greatly contribute to national and international long-term and large scale research, education, and service missions as we enter the 21st century.

Acknowledgements

This chapter represents an overview and culmination of 25 years of research in the areas of agroecosystem and landscape ecology at the Ecology Research Center, Miami University, Oxford, Ohio, U.S.A. There is always the risk of omitting names of colleagues, as well as funding agencies, who have made this long-term research possible. However, I wish to especially thank Post-Doctoral Scholars Eric K. Bollinger, David M. Pavuk, John D. Peles, and Nicholas L. Rodenhouse; as well as graduate students Patrick J. Bohlen, Mary Benninger-Traux, Rachael J. Collins, Steven J. Harper, Janice C. Kemp, Vincent N. La Polla and Gregory C. Lorenz, who interacted with me to make this research both exciting and worthwhile. Numerous undergraduates, especially Susan R. Brewer, Megan Casey, Dwight M. Holmes, Michael F. Lucas, Christopher K. Williams and Valerie A. Whitman, also played a vital role in these agroecosystem and agrolandscape level investigations. These studies were funded by NSF Grants DEB-8004176, NSF BSR-9006451, and NSF BSR-8818086; EPA Grants R-812385-01 and EPA R-815033-01; and USDA grant 9100824. Earthwatch volunteers also participated in these investigations from 1986 through 1988. Last, but not least, I thank Barbara Ekbom for the invitation to contribute this chapter and for her professional interactions regarding this important endeavor.

5.6 References

Allen, F.H. & Hoekstra, T.W., 1992. Toward a Unified Ecology. Columbia University Press, New York, NY.

Altieri, M.A., Schoonhoven, A. & Doll, J.D., 1977. The ecological role of weeds in insect pest management systems: A review illustrated with bean (*Phaseolus vulgaris*) cropping systems. *Proceedings of the National Academy of Sciences (USA)* **23**:185-205.

Baldwin, A.D., de Luce, J. & Pletsch, C. (Eds.), 1994. *Beyond Preservation: Restoring and Inventing Landscapes*. University of Minnesota Press, Minneapolis, MN.

Barrett, G.W., 1968. The effects of an acute insecticide stress on a semi-enclosed grassland ecosystem. *Ecology* **49**: 1019-1035.

Barrett, G.W., 1985. A problem-solving approach to resource management. *BioScience* **35**:423-427.

Barrett, G.W., 1989. Viewpoint: A sustainable society. *BioScience* **39**: 754.

Barrett, G.W., Rodenhouse, N.L. & . Bohlen, P.J, 1990. Role of sustainable agriculture in rural landscapes. In: Edwards, C.A., Lal, R., Madden, P., Miller, R.H. & House, G., (Eds.), *Sustainable Agricultural Systems.* Soil and Water Conservation Society, Ankeny, IA, pp. 624-636.

Barrett, G.W. & Bohlen, P.J., 1991. Landscape ecology. In: Hudson, W.E. (Ed.), *Landscape Linkages and Biodiversity.* Island Press, Washington, D.C., pp. 149-161.

Barrett, G.W. 1992. Landscape ecology: Designing sustainable agriculture landscapes. In:. Olsen, R.K., (Ed.), *Integrating Sustainable Agriculture, Ecology, and Environmental Policy.* Haworth Press, Inc., New York, NY, pp. 83-103.

Barrett, G.W. 1994. Restoration ecology: Lessons yet to be learned. In: Baldwin, D., de Luce, J. & Pletsch, C. (Eds.), *Preservation: Restoring and Inverting Landscapes.* University of Minnesota Press, Minneapolis, MN, pp. 113-126.

Barrett, G.W. & Peles, J.D., 1994. Optimizing habitat fragmentation: An agrolandscape perspective. *Landscape and Urban Planning* **28**: 99-105.

Barrett, G.W., Peles, J.D. & Harper, S.J., 1995. Reflections on the use of experimental landscapes in mammalian ecology. In: Lidicker, Jr., W.Z. (Ed.), *Landscape Approaches in Mammalian Ecology and Conservation.* University of Minnesota Press, Minneapolis, MN, pp. 157-174.

Barrett, G.W., Peles, J.D. & Odum, E.P., 1997. Transcending processes and the levels-of-organization concept. *BioScience* **47**:531-535.

Barrett, G.W. & Peles, J.D., 1999. *Landscape Ecology of Small Mammals.* Springer-Verlag, New York, NY.

Bohlen, P.J. & Barrett, G.W., 1990. Dispersal of the Japanese beetle, *Popillia japonica* (Coleoptera, Scarabaeidae), in strip-cropped soybean agroecosystems. *Environmental Entomology* **19**: 955-960.

Callahan, J.T., 1984. Long-term ecological research. *BioScience* **34**: 363-367.

Carpenter, S.R. & Kitchell, J.F., 1988. Consumer control of lake productivity. *BioScience* **34**: 764-769.

Carpenter, S.R. & Kitchell, J.F., (Eds.), 1993. The Trophic Cascade in Lakes. Cambridge University Press, Cambridge, England.

Carson, W.P. & Barrett, G.W., 1988. Succession in old-field plant communities: Effects of contrasting types of nutrient enrichment. *Ecology* **69**: 984-994.

Diffendorfer, J.E., Slade, N.A., Gaines, M.S. & Holt, R.D.,. 1995. Population dynamics of small mammals in fragmented and continuous old-field habitat. In: Lidicker, Jr., W.Z. (Ed.), *Landscape Approaches in Mammalian Ecology and Conservation.* University of Minnesota Press, Minneapolis, MN, pp. 175-199.

Ehrlich, P.R. & Murphy, D.D., 1987. Monitoring populations on remnants of native habitat. In: Saunders, D. & Hopkins, A., (Eds.), *Nature Conservation: The Role of Remnants of Native Vegetation.* Hobbs, Surrey, Beatty & Sons, Australia, pp. 201-210.

El Titi, A. & Landes, H., 1990. Integrated farming system of Lautenbach: A practical contribution toward sustainable agriculture in Europe. In: Edwards, C.A.,Lal, R., Madden, P., Miller, R.H. & House, G. (Eds.), *Sustainable Agriculture Systems.* Soil and Water Conservation Society, Ankeny, IA, pp. 265-286.

Fahrig, L. & Paloheimo, J., 1988. Effect of spatial arrangement of habitat patches on local population size. *Ecology* **69**: 468-475.

Forman, R.T.T. & Godron, M., 1986. Animal and plant movement across a landscape. In: Forman, R.T.T. & Godron, M., (Eds.), *Landscape Ecology.* John Wiley and Sons, New York, NY, pp. 357-395.

Forman, R.T.T. & Baundry, J., 1984. Hedgerows and hedgerow networks in landscape ecology. *Environmental Management* **8**: 495-510.

Goodland, R., 1995. The concept of environmental sustainablity. *Annual Review of Ecology and Systematics*

26: 1-24.

Holmes, D.M. & Barrett, G.W., 1997. Japanese beetle (*Popillia japonica*) dispersal behavior in intercropped vs. monoculture soybean agroecosystems. *American Midland Naturalist* **137**: 312-319.

Hurlbert, S.H., 1984. Pseudoreplication and the design of ecological field experiments. *Ecological Monographs* **54**: 187-211.

Jantsch, E., 1972. *Technological Planning and Social Futures*. Cassell Associated Business Programmes, London, England.

Johnson, P.L., (Ed.), 1977. *An Ecosystem Paradigm for Ecology*. Oak Ridge Associated Universities, Oak Ridge, TN.

Kemp, J.C. & . Barrett, G.W., 1989. Spatial patterning: Impact of uncultivated corridors on arthropod populations within soybean agroecosystems. *Ecology* **70**: 114-128.

Kogan, M. & Kuhlman, D.E., 1982. Soybean insects: Identification and management in Illinois. Bulletin 773, College of Agriculture, University of Illinois, Urbana-Champaign, IL.

La Polla, V.N. & Barrett, G.W., 1993. Effects of corridor width and presence on the population dynamics of the meadow vole (*Microtus pennsylvanicus*). *Landscape Ecology* **8**: 25-37.

Likens, G.E., 1989. *Long-term Studies in Ecology: Approaches and Alternatives*. Springer-Verlag, New York, NY.

Lorenz, G.C. & Barrett, G.W., 1990. Influence of simulated landscape corridors on house mouse (*Mus musculus*) dispersal. *American Midland Naturalist* **123**: 348-356.

Lubchenco, J., Olson, A.M., Brubaker, L.B., Carpenter, S.R., Holland, M.M, Hubbell, S.P., Levin, S.A., MacMahon, J.A., Matson, P.A., Melillo, J.M., Mooney, H.A., Peterson,C.H., Pulliam, H.R., Real, L.A., Regal, P.J. & Risser, P.G., 1991. The Sustainable Biosphere Initiative: An ecological research agenda. *Ecology* **72**: 371-412.

MacMahon, J.A., Phillips, D.L., Robinson, J.V. & Schimpf, D.J., 1978. Levels of biological organization: An organism-centered approach. *BioScience* **28**: 700-704.

National Research Council, 1989. *Alternative Agriculture*. National Academy Press, Washington, D.C..

New, T.R., 1991. *Butterfly Conservation*. Oxford University Press, Oxford, England.

Nichols, A.O. & Margules, C.R., 1991. The design of studies to demonstrate the biological importance of corridors. In: Saunders, D.A. & Hobbs, R.J. (Eds.), *Nature Conservation 2: The Role of Corridors*. Hobbs, Surrey, Beatty & Sons, Australia, pp.49-64.

Noss, R.F., 1983. A regional landscape approach to maintain diversity. *BioScience* **33**: 700-706.

Noss, R.F., 1991. Landscape connectivity: Different functions at different scales. In: Hudson, W.E. (Ed.), *Landscape Linkages and Biodiversity*. Island Press, Washington, D.C., pp. 27-39.

Odum, E.P., 1969. The strategy of ecosystem development. *Science* **164**: 262-270.

Odum, E.P., 1997. *Ecology: A Bridge between Science and Society*. Sinauer Associates, Inc., Sunderland, MA.

Neill, R.V., De Angelis, D.L., Waide, J.B. & Allen, T.F.H., 1986. *A Hierarchical Concept of Ecosystems*. Princeton University Press, Princeton, NJ.

Pastor, J., 1995. Ecosystem management, ecological risk, and public policy. *BioScience* **45**: 286-288.

Pavuk, D.M. & Barrett, G.W., 1993. Influence of successional and grassy corridors on parasitism of *Plathypena scabra* (F.) (Lepidoptera: Noctuidae) larvae in soybean agroecosystems. *Environmental Entomology* **22**: 541-546.

Peles, J.D., Brewer, S.R. & Barrett, G.W., 1996. Metal uptake by agricultural plant species grown in sludge-amended soil following ecosystem restoration practices. *Bulletin Environmental Contamination and Toxicology* **57**: 917-923.

Pulliam, H.R., Dunning, Jr., J.B. & Liu, J., 1992. Population dynamics in complex landscapes: A case study. *Ecological Applications* **2**: 165-177.

Reaka-Kudla, M.L., Wilson, D.E. & Wilson, E.O., (Eds.), 1997. Biodiversity II. National Academy Press, Washington, D.C.

Risser, P.G., Karr, J.R. & Forman, R.T.T., 1984. Landscape ecology: Directions and approaches. *Illinois Natural History Survey, Special Publication*. Champaign, IL. **2**: 18 pp.

Rodenhouse, N.L., Barrett, G.W., Zimmerman, D.M. & Kemp, J.C., 1992. Effects of uncultivated corridors on arthropod abundances and crop yields in soybean agroecosystems. *Agriculture, Ecosystems and Environment* **38**: 179-191.

Rowe, J.S., 1961. The level-of-integration concept in ecology. *Ecology* **42**: 420-427.

Salwasser, H., 1991. Roles and approaches of the USDA Forest Service. In: Hudson, W.E., (Ed.), *Landscape Linkages and Biodiversity*. Island Press, Washington, D.C., pp. 54-65.

Schoonhoven, A.V., Ardona, C., Garcia, J. & Garzon, F., 1981. Effect of weed covers on *Empoasca kraemeri* populations and dry bean yields. *Environmental Entomology* **10**: 901-907.

Seastedt, T., 1996. Ecosystem science and society. *BioScience* **46**: 370-372.

Shelterbelt Project. 1934. (Published statements by numerous separate authors.) *Journal of Forestry* **32**: 952-991.

Stinner, R.E., Barfield, C.S., Stimac, J.L. & Dohse, L., 1983. Dispersal and movement of insect pests. *Annual Review of Entomology* **28**: 319-335.

Williams, C. K., Witmer, V.A., Casey, M. & Barrett, G.W., 1994. Effects of strip-cropping on small mammal population dynamics in soybean agroecosystems. *Ohio Journal of Science* **94**: 94-98.

CHAPTER 6

INTERCHANGES OF A COMMON PEST GUILD BETWEEN ORCHARDS AND THE SURROUNDING ECOSYSTEMS

A Multivariate Analysis of Landscape Influence

PHILIPPE JEANNERET
Swiss Federal Research Station for Agroecology and Agriculture, Zürich, Switzerland

6.1 Introduction

In orchards, insect pests are increasingly controlled using integrated pest management (IPM) (Blommers, 1994). Among insect pests, tortricids (Lepidoptera: Tortricidae) are frequent all around the world and sometimes are very abundant (*e.g.* Dickler, 1991), including the codling moth (*Cydia pomonella* L.), the summer fruit tortrix (*Adoxophyes orana* F.v.R.), the eyespotted budmoth (*Spilonota ocellana* F.), *Pandemis heparana* Den. & Schiff., the green budmoth (*Hedya nubiferana* Haw.), the European leafroller (*Archips rosana* L.), the fruit tree tortrix (*A. podana* Scop.), the brown oak tortrix (*A. xylosteana* L.), *A. crataegana* Hbn., and the small fruit tortrix (*Grapholita lobarzewskii* Nowicki). The bionomics of these species are usually very well known (Balachowsky, 1966; Chambon, 1986; Dickler, 1991). Distribution and movement patterns into and from the orchard have been investigated with marking-recapture methods, using pheromone traps or alimentary baits (Mani & Wildbolz, 1977; Sziraki, 1984; Brunner *et al.*, 1988; Fassotte *et al.*, 1992). Similarly, movements between agroecosystems and their surroundings have been studied for surface-dwelling carabids, spiders, etc. and aerial plankton (Altieri & Schmidt, 1986; Duelli *et al.*, 1989; Duelli *et al.*, 1990).

Interactions between ecosystems are an important topic from an ecological point of view (Gulinck, 1986; Baudry, 1989; Burel, 1992; Dennis & Fry, 1992; Burel & Baudry, 1995) and need to be quantified (Stinner *et al.*, 1983). Measured global process (global exchanges between ecosystems) are characterized by their multidimensional aspect, because all variables (each and every species) are considered together.

On the Swiss plateau, managed apple orchards are either use integrated pest management (IPM) or chemical control. Sometimes orchards are abandoned, and, in the Lemanic Basin, high trees which are pruned once a year are only of interest to the owners for the meadow, used in spring for cattle grazing and in summer for hay or green fodder.

The first hypothesis of this study was to suppose the existence of important environmental factors on the tortricid activity at the margin of apple orchards. In measuring this influence as well as its statistical meaning we will be able to draw conclusions about the ecological role of the orchard for tortricids in the surrounding landscape. As a starting point, we considered that the direct environment of the orchard was the main influencing

B. Ekbom, M. Irwin and Y. Robert (eds.), Interchanges of Insects, 85-107

factor. In fact, we could not imagine that fauna living in an orchard surrounded by forest could not be influenced by this forest. This remark is particularly important in the case of mobile insects. We can also say that the orchard is very often visited by forest species and that typical orchard species also explore the forest in a regular manner. Nevertheless, these numerous exchanges do not mean that these species fix themselves in the adjacent environment.

This paper presents the results of study using trapping of moths at the margin of three IPM and three abandoned orchards in the area Nyon-La Côte (Lemanic Basin), Switzerland. Data on emigrant and immigrant tortricids, caught with a Malaise trap were examined using multivariate analysis as a whole and for those species that are associated with apple trees. Analysis procedures take as an example data that concern associated species.

6.2 Field Methods and Sites

Moths were caught using a transformed bi-directional Malaise trap (Townes, 1972) with a 1 m x 2 m interception screen (1 mm^2 mesh). The trap was 3 m high and caught insects between 1-2 m height, which represents the apple tree crown height. Two collectors were placed on top of an aluminium frame to separate emigrant from immigrant individuals. Each collector contained 250 ml alcohol (75%).

A Malaise trap was placed on each of the North, East, South and West margins of six apple orchards. From the beginning of July to October, 48 samples were collected on 14 successive weeks. Tortricids were isolated from all other insects in each sample and species were determined by their genitalia based on the criteria of Hannemann (1961), Graff Bentinck and Diakonoff (1968), Chambon (1986), and Kuznetsov (1989).

Abandoned orchards were not pruned and comprised 15 to 25 trees. In IPM orchards (1000 to 4000 trees) the mating disruption technique was applied against codling moth and no insecticides had been used for 10 years. In one orchard, virus and growth regulators were also used.

6.3 Statistical Methods

First of all, applied techniques belong to the classical unidimensional statistic field. (Fig. 1). A substantial part of data is in the form of means and standard deviations (descriptive statistics), calculated on number of individuals and number of species. In this part of the study, inferential statistical treatment of the data is possible to use, like the Wilcoxon-Mann-Whitney U test and the Wilcoxon test for matched samples, when no self-correlation is present (Scherrer, 1984).

Secondly, analysis always simultaneously considers the species on the whole and their respective abundance (Fig. 2). We speak in this case of multivariate analysis as well as multidimensional statistics.

Practical applications of the statistical analysis are very closely related to the use of specific software. The following software was used for the multivariate analysis:

- CANOCO© (Microcomputer Power, Ithaca) and Ter Braak (1986, 1987a, 1990a, 1991, 1992)
- PROGICIEL® (Montreal University) and Legendre & Vaudor (1991).

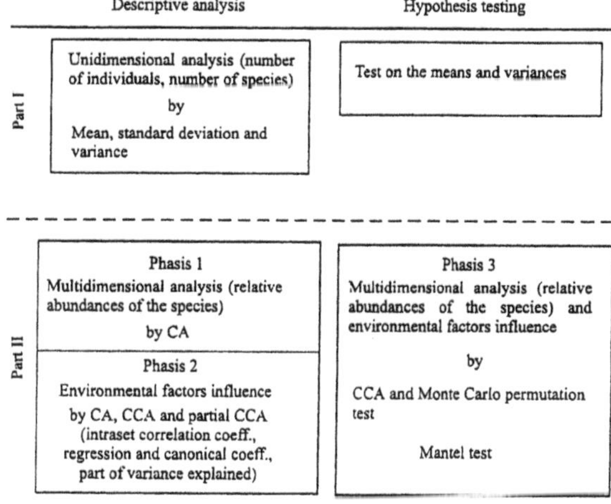

Figure 1. Model used to describe and analyse trends in community and species-environment relationship.

Three steps characterise this second part of the analysis:

1) Exploratory stage: ordination diagrams obtained with correspondence analysis (CA) and overlapping of results of one clustering. In our study, ordinations give a very good picture of similarities and dissimilarities between objects, but we will also use clustering to confirm some observed structures and to sharpen relations between these objects.
2) Data interpretative stage, in three steps:
   - Indirect analysis in which we try to explain subsequent structures of the explanatory stage obtained through a CA with the help of explanatory factors (environmental descriptors). Technically, we used intraset correlation coefficients (correlation between explanatory factors and the ordination axis, defined by Ter Braak, 1986, p. 1170), as well as multiple regression coefficients of site coordinates (or surveyed sites) on the explanatory factors (regression is calculated after having extracted the species and site coordinates from the CA).
   - Direct analysis in which structures are revealed introducing explanatory factors. In our study this is done through a canonical correspondence analysis (CCA). We then

use intraset correlation coefficients as well as multiple regression canonical coefficients of the environmental axis (site coordinates obtained through a linear combination of explanatory factors) on the data.

- Partition of variance in which we extract the significant share of each of the introduced explanatory factors. This analysis is realised by means of a series of partial CCA.

3) Testing of hypothesis: statistical tests that allow confirmation of the hypothesis of a significant action of the explanatory factors on the data structure. The applied procedure is a permutation test, so-called Monte Carlo, and is realised during the CCA (Hope, 1968).

The detailed model (Fig. 2) shows how to use CCA with each factor, separately and then through forward selection, followed by CCA with all factors taken together. The goal is to establish a hierarchy between factors and to eliminate the ones which do not explain any variance.

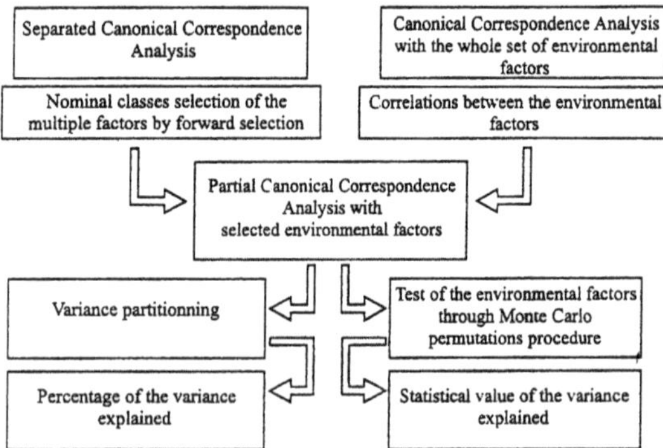

Figure 2. Detailed model used to describe and analyse trends in community and species-environment relationship with multivariate statistics.

Partitioning of variance is then performed through partial CCA. The fraction of the variance explained (and its significance, Monte Carlo permutation test) by each of the environmental descriptors is given separately, after eliminating the variance due to the other (partialed) factors, which are used as covariables.

Rough data have been transformed through the y' = ln(y+1) function. This transformation is particularly well adapted to species abundance. This type of data is generally made up of a great number of species with few individuals and some abundant species throughout all the sampling sites. Such a distribution does not correspond to a normal distribution. Logarithmic transformation allows us to moderate the impact of a few very abundant species on the analysis (Legendre & Legendre, 1984, p. 18-19; Jongman *et al.*, 1987, p. 103). These authors recommend disregarding species with very few individuals. In our study, our limit has been set to 5 individuals.

6.4 Results

6.4.1 DESCRIPTIVE ANALYSIS, PART I

A total of 1543 (1992) and 1377 (1993) tortricids were caught, representing 88 (1992) and 95 (1993) species, 430 (1992) and 325 (1993) from IPM orchards, 1113 (1992) and 1052 (1993) from abandoned orchards. An important difference was observed in the number of species caught between IPM (44 and 39, in 1992 and 1993 respectively) and abandoned orchards (81 and 65), immigration being usually more important than emigration except for two IPM orchards (Fig. 3).

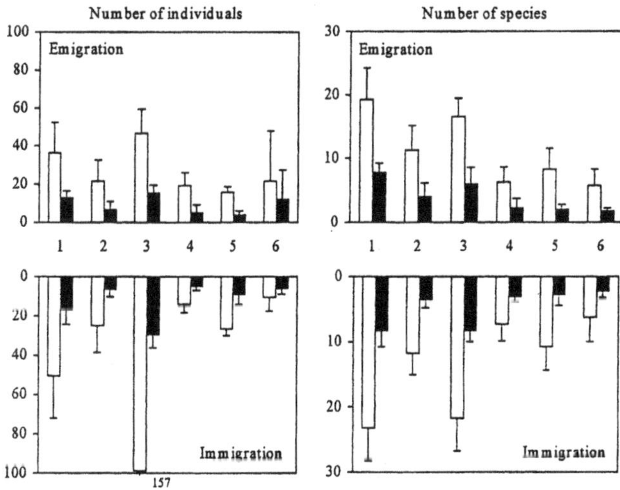

Figure 3. Mean (and standard deviation) number of emigrant and immigrant tortricids at six apple orchards calculated on the basis of four Malaise traps (1992). White bars represent the overall community and black bars the associated species. Uncultivated orchards: 1, 2, 3. IPM orchards: 4, 5, 6.

Standard deviations are important and reflect heterogeneity between orientation. (*i.e.* the Malaise traps). The number of species emigrating and immigrating is similar. Species associated with apple trees show the same trend of individuals and species variation as the whole data set. The proportions of associated species (26 to 48%) and individuals (26 to 34%) was similar in each orchard. The general patterns described above are repeated in the 1993 data (not presented).

6.4.2 DESCRIPTIVE ANALYSIS, PART II, PHASES 1 AND 2

Ordination diagrams which represent the CA results done with the means of a reduced species matrix, *i.e.* only with the abundance of species associated to the apple tree (associated species guild), differentiate the 24 interfaces in a very good manner. In fact, axis have high eigenvalues, which clearly shows that species are well differentiated (Table 1). Malaise traps in orchard 1 and 3 are relatively close together (Figs. 4, 5). We can also observe on the upper part of this diagram a homogeneous group which is made up of a majority of Malaise traps set at the interface of crops and adjacent ecosystems.

Species separation is important all along axis 1 (high eigenvalue of this axis) and corresponds to the traditional orchard separation. We can stress that *Sparganothis pilleriana* (SPPI) is often found in orchard 2, *Archips crataegana* (ARCR), *Batodes angustiorana* (BAAN), *Pandemis cerasana* (PACE) *Archips xylosteana* (ARXY), *Acleris rhombana* (ACRH) and *Archips podana* (ARPO) are often found in orchard 3. Orchard crops are not always characterized by one or many species in particular but *Pandemis heparana* (PAHE) is almost found everywhere.

Table 1. CA 1992 for the associated species guild. Analysis is done by means of a 17 species matrix, distributed on 24 Malaise traps

	1992				
Axis	1	2	3	4	
Eigenvalues	0.397	0.273	0.165	0.104	
Total inertia = sum of the unconstrained eigenvalues					1.557

6.4.3 DESCRIPTIVE ANALYSIS, PART II, PHASES 2 AND 3

Primary influencing factors were defined and introduced in a canonical correspondence analysis (CCA) to understand their impact on the tortricid population at the interface between the orchard and its surroundings. At first, we selected a number of descriptors. After that, we did a series of partial CCA to determine how much of the total variation could be explained by the environment. This procedure is explained in Fig. 2. Influence of descriptors was then tested using the CCA axis to confirm results obtained through the CCA (Ter Braak, 1986).

CCA allows us to test the hypothesis of a species/environment relation. In our case, hypothesis was "active tortricids at the interface between a orchard/crop and the neighbouring environment are significantly influenced by: the type of surrounding environment, the further environment of the orchard, the orchard type (traditional versus cultivated orchards), the orientation of the interface as well as the geographical position of the orchard". The procedure used in CANOCO program is the Monte Carlo test (permutation test).

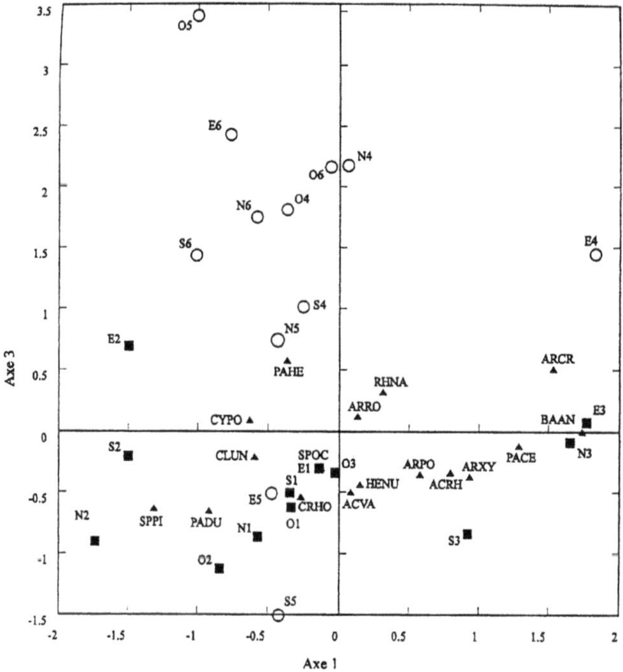

Figure 4. Detrended CA which presents apple tree associated species tortricids caught in 24 Malaise traps set at the margins of 6 orchards and crops. Species which count less than 5 individuals are not taken into account. 1, 2, 3: orchards, 4, 5, 6: crops. N, E, S,O: Malaise trap position, *i.e.* North, East, South or West. Species description are made up first letter of genus and species.

We hereby present the set of basic descriptors which we have defined "a priori" (Table 2). The 8 chosen descriptors characterise the orchard environment or the Malaise trap environment. Four of them are nominal, two fix the orchard coordinates, one is quantitative and the last one is semi-quantitative.

The first descriptor, which defines the orchard type is divided in two classes, either crop (v1) or orchard (v2). This is a nominal and explanatory variable.

We could suppose that the landscape would act on two different levels on the interface activity. The primary action circle would be given by the immediate surroundings of the orchard (immediately adjacent ecosystem), which we will call "direct landscape effect" as it is the result of a direct contact between ecosystems. The descriptor which defines the type of ecosystem which is directly adjacent to the orchard (nominal descriptor) is divided in 6 classes, *forest* (v3) *intensive crops* (v4), *vineyards* (v5), *other orchard* (v6), *meadow* (v7), and *garden* (v8). This explanatory variable which refers to the direct impact of the landscape on the tortricid activity at the interface has been called *landscape I*.

Table 2. Multiclass environmental descriptors used as explanatory factors in the multivariate analysis of the tortricid activity at the interface of orchard/crops and their adjacent ecosystems

Environmental descriptors		Classes of environmental descriptors	
Orchard type		v1	Crop
		v2	Orchard
Type of adjacent surroundings		v3	Forest I
		v4	Intensive crop I
		v5	Vineyard I
Landscape I		v6	Other Orchard I
		v7	Meadow I
		v8	Garden I
Type of further surroundings		v9	Forest II
		v10	Intensive crop II
		v11	Vineyard II
Landscape II		v12	Other Orchard II
		v13	Meadow II
		v14	Garden II
Space	**Position**	v15	West
		v16	North
		v17	East
		v18	South
	Geographical coordinates	v19	x
		v20	y
Border/area		v12	quantitative
Structure		v22	semi-quantitative
Emmigration-Immigration		v23	emigration
		v24	immigration

The second action circle is the one which is further from the orchard (300 m away from the interface) and is called indirect landscape effect. It is a nominal descriptor and made up of 6 classes identical to the ones used in the description of *landscape I* (v9 to v14). It measures the regional impact of the landscape on the tortricid activity. This explanatory variable is very logically called *landscape II*.

The fourth descriptor, also nominal, indicates the *orientation* of the Malaise trap according to the 4 cardinal points West(v15), North (v16), East (v17) and South (v18). It also describes the possible movements of tortricids on a long distance basis, from east to west or from north to south, for example. In the La Côte region, the north-south direction cuts at a perpendicular the Jura mountain axis and the Leman lake and is characterised by a very typical wind, the "Joran", which blows mainly in summer. The East-South axis represents the prevailing direction of winds in the region.

The fifth descriptor is defined by the *geographical coordinates* of the orchards which were measured on the basis of a topographical map (scale 1/25.000) (v19 and v20).

The sixth descriptor indicates the relation between the length of border (of the orchard) and the sire of its area (quantitative descriptor) and is called *border/area* (v21).

The seventh descriptor describes the *structure* (v22) of the adjacent surroundings according to its openness (semi-qualitative descriptor, scale 1 to 3). For example, a building is considered as a barrier and will give a value of 1 for this element, as well as a forest edge.

We considered that the influence of an environmental descriptor is the same on the two collectors set on one Malaise trap. The multiple classes of the 8 environmental descriptors are considered in the analysis as v1 to v22 explanatory variables.

Concerning codes for sites or collectors, we can underline that the first letter gives the orientation, that is W for West, N for North, E for East and S for South. The following digit indicates the orchard's number and the last letter the collectors position, *i.e.* E for emigration and I for immigration. Values of nominal variables are binary either 0 or 1. Each time a class characterises a border of an orchard, the value is 1. For example, north and east sides of orchard 3 are bordered by forest. The title of each variable will from now on be written in cursive letters (example: *landscape I*).

CCA with the whole set of environmental descriptors

First objective of this analysis is to eliminate environmental descriptors which are highly correlated to each other. Without giving details of eigenvalues related to each axis, 60.5% of total inertia for the guild matrix is explained by the 24 environmental descriptors used. These results are similar to the ones obtained for the whole population (results not presented). This means that the apple tree associated tortricid guild behavior in relation to the environmental descriptors must be about the same. In fact, if the guild would have a completely different behavior, we could not, for example, explain such a high percentage of variation with the same environmental descriptors. The *edge/area* and the *structure* relation were descriptors largely explained by other descriptors (Pearson's $r > 0.5$). We decided then to take the descriptors, which explain a more important part of the variation (*i.e. landscape I* instead of *edge/area* and *structure*).

CCA descriptor by descriptor and forward selection of the multiclass environmental descriptors

It is interesting to study the behavior of the apple tree associated tortricid guild in relation to the environmental descriptors to enable us to find out what part the landscape plays inrelation to the guild activity. We have therefore conducted a series of separated CCA for each of the descriptors (Table 3) and this followed with a forward selection of the landscape descriptor's classes (*landscape I* and *landscape II*; Table 4).

Table 3. Separated CCA. This analysis is done with the means of a 17 species matrix distributed on 48 collectors set two on 24 Malaise traps which were installed at the margin of 6 orchard/crops. Axis are under constraint of the following environmental descriptors: movement direction (*emigration-immigration*), *orientation* (west, north, east and south), *geographical coordinates* and *orchard type*. Monte Carlo test was conducted on canonical axis with 99 permutations. Significant values on a 1% level are written in bold letters.

	Emigration - Immigration				
Axis	1	2	3	4	Total Inertia
Eigenvalues	0.075	0.407	0.394	0.296	2.532
Sum of canonical eigenvalues					0.075
Percentage of variation					3
Monte Carlo test					*p = 0.14*

	Orientation				
Axis	1	2	3	4	Total Inertia
Eigenvalues	0.121	0.046	0.021	0.433	2.532
Sum of canonical eigenvalues					0.188
Percentage of variation					7.4
Monte Carlo test					*p = 0.23*

	Geographical coordinates				
Axis	1	2	3	4	Total Inertia
Eigenvalues	0.276	0.071	0.397	0.306	2.532
Sum of canonical eigenvalues					0.348
Percentage of variation					13.7
Monte Carlo test					***p = 0.14***

	Orchard type				
Axis	1	2	3	4	Total Inertia
Eigenvalues	0.182	0.432	0.381	0.281	2.532
Sum of canonical eigenvalues					0.182
Percentage of variation					7.2
Monte Carlo test					***p = 0.001***

Table 4. Separated CCA. This analysis is done with the means of a 17 species matrix distributed on 48 collectors set two on 24 Malaise traps which were installed at the margin of 6 orchard/crops. Axis are under constraint of 2 environmental descriptors representing the landscape at two different levels (*landscape I and landscape I*). CCA is followed by a forward selection of descriptors's classes. Descriptors are numbered following their coming out during selection. Monte Carlo test was conducted on canonical axis with 99 permutations.

	Landscape I				
Axis	1	2	3	4	Total Inertia
Eigenvalues	0.247	0.154	0.053	0.039	2.532
Sum of canonical eigenvalues					0.497
Percentage of variation					19.6
Forward selection	1 Forest I				
	Values after introduction of selected variables				
Axis	1	2	3	4	Total Inertia
Eigenvalues	0.230	0.435	0.297	0.260	2.532
Sum of canonical eigenvalues					0.230
Percentage of variation					9.1
	Landscape II				
Axis	1	2	3	4	Total Inertia
Eigenvalues	0.258	0.127	0.088	0.046	2.532
Sum of canonical eigenvalues					0.555
Percentage of variation					21.9
Forward selection	1 Forest II				
	Values after introduction of selected variables				
Axis	1	2	3	4	Total Inertia
Eigenvalues	0.233	0.400	0.297	0.263	2.532
Sum of canonical eigenvalues					0.233
Percentage of variation					9.2

The way the collectors were positioned, catching immigration or emigration fauna has no influence on this scale (Table 3). For a given interface emigration and immigration of apple tree associated totricids are statistically the same, the *emigration - immigration* descriptor does not explain a significant share of the guild matrix ($p = 0.14$). In the same way, Malaise trap *orientation* does not explain a significant share of the species matrix ($p = 0.23$). On the other side, *geographical coordinates* have an influence on the activity distribution, as the descriptor explains 13.7% of variation, as well as the *orchard type* (7.2% of the variation explained).

In the case of landscape multiple classes descriptors, a first CCA calculates the eigenvalues for the axis and the percentage of variation explained by all the multiple classes taken together and later selects the classes, in the case where they explain a significant share of the reminding variation. A second CCA calculates the new eigenvalues and the new percentage of variation explained by the selected classes. As before, we used the Monte Carlo test to statistically evaluate the influence of each class (level 1%). A significant part of the species matrix is explained by the *landscape I* and the *landscape II* through their class *forest* (Table 4).

We can here stress that in the interpretation of results, choice of significance level is very important. It is very often important to have the real value of p to be able to analyse the results, because values can be very close to significance level. P-values obtained during permutation test, leading to the elimination of environmental descriptors such as *orientation* or movement direction were significantly higher and could not be used in the analysis, even taking a significance level at 5%. On the contrary, if we would have taken this significance level of 5%, as sometimes done in an ecological study, results of the successive selections of the *landscape I* and *landscape II* descriptors would be modified. Especially, with a 5% significance level, the *vineyard I* and *II* descriptors would be selected in 1992, as well as *forest I* in 1993 (Table 5).

The variability of the associated guild activity is primarily influenced by the *forest* as an adjacent ecosystem, but *vineyard* is also important.

Table 5. Summary of p-values obtained in the forward selection of the landscape descriptors (landscape I et landscape II) on the matrix guild

	Landscape I		Landscape II	
Classes	1992	1993	1992	1993
Forest	$p =$	$p =$	$p =$	$p > 0.1$
Intensive crops	$p > 0.1$	$p > 0.1$	$p = 0.04$	$p = 0.08$
Vineyard	$p =$	$p > 0.1$	$p > 0.1$	$p > 0.1$
Orchard	$p > 0.1$	$p > 0.1$	$p > 0.1$	$p > 0.1$
Meadow	$p > 0.1$	$p > 0.1$	$p > 0.1$	$p > 0.1$
Garden	$p > 0.1$	$p > 0.1$	$p > 0.1$	$p > 0.1$

CCA ordination diagrams axis interpretation and comparison with the CA results

CCA obtained on the basis of the apple tree associated species has been realised for many of the 17 species and constrained by the following 4 environmental descriptors, which were selected after examination of correlation between descriptors and forward selection: *orchard type, geographical coordinates, landscape I* (one and only class = *forest I*) end *landscape II* (one and only class = *forest II*).

Malaise traps which were set at the margin of the orchards 1 and 3 are very close and very near to the descriptors *forest II* and traditional orchard (Fig. 6) Malaise traps which

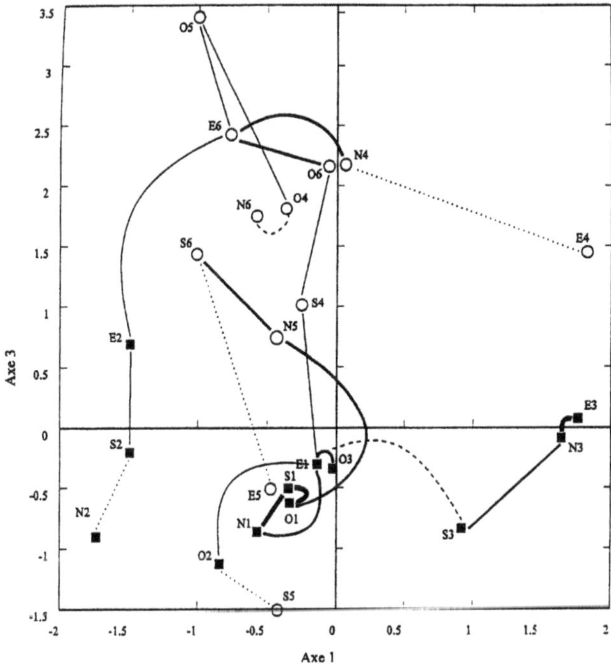

Figure 5. Superimposition of the shortest spanning tree (single linkage clustering) to the CA (axis 1 and 3) of the Fig. 4. Thick line: distance < 0.25; Medium line: 0.25 ≤ distance < 0.35; Thin line: 0.35 ≤ distance < 0.42; broken line: 0.42 ≤ distance < 0.48; dotted line: 0.48 ≤ distance ≤ 0.6.

are set near the cultivated orchards are very logically closer to this descriptor. We can point out the shape of a group that assembles at a distance of less than 0.5 for trap numbers O6, E6, S4 and N4. If we study the detail of the results of the single linkage clustering, we can deduce that the traps grouped in orchard number 2 are related to the group of traps set in crops and at a distance of 0.64.

If we study eigenvalues and correlation between species-environment in both CCA and CA, we may conclude that we see a typical case, because the eigenvalues of the CCA are slightly weaker and the correlation factors higher (Table 6). The first axis of the CA gives full information. Nevertheless, the axis 2 of the CCA reveals in a very precise way the species-environment relationship.

Table 6. CA and CCA. The analysis is done on the basis of a 17 species matrix caught in 24 Malaise traps set at the margin of 6 orchard/crops. CCA axis are under constraint of 4 environmental descriptors divided in 6 classes. Eigenvalues and correlation coefficients between the species-axis and the environmental-axis for the 4 first axis (species-environment correlation)

	Axis			
	1	2	3	4
	Eigenvalues			
CA	0.397	0.273	0.165	0.104
CCA	0.344	0.225	0.093	0.074
	Species-environment correlations			
CA	0.925	0.821	0.741	0.311
CCA	0.936	0.942	0.757	0.697
Total inertia				1.557
Sum of canonical eigenvalues				0.751

If we consider the correlations between environmental descriptors, we notice very weak links between them. In fact, the greatest correlation coefficient is 0.53 (if we omit the *latitude-longitude* correlation). It is then possible to give an explanation to the canonical coefficients and this explanation should be the same as the one used for the intraset correlations.

Table 7. Intraset correlation foor the CCA and the CA. Coefficient for the 3 main descriptors are written in bold

	Intraset correlation coefficients							
	CA				CCA			
Axis	1	2	3	4	1	2	3	4
Descriptors								
Orchard type	-0.21	**0.55**	**0.93**	**0.39**	-0.25	**0.73**	**0.60**	-0.04
Forest I	0.67	**0.58**	0.08	-0.06	0.67	**0.49**	**-0.45**	**0.31**
Forest II	**0.80**	-0.07	-0.08	**0.22**	**0.80**	-0.06	0.08	**0.25**
Longitude	**-0.75**	0.21	**0.28**	**0.50**	**-0.74**	**0.22**	-0.04	**0.63**
Latitude	**-0.90**	**0.22**	-0.02	0.07	-0.89	0.18	**-0.21**	0.21

Intraset correlations of CA and CCA give homogeneous results, axis 1 is a geographical axis because *latitude* is the highest value. On this axis, Malaise traps which were set at the far East (*longitude*) are also the ones which have the least forest in the surroundings. Axis 2 is fixed by *forest I* and *orchard type*. Axis 3 is also an axis *orchard type* and axis 4 is a longitudinal axis.

Canonical and regression coefficients both indicate that *latitude* is determinant on axis 1. For these coefficients, axis 2 is also latitudinal, but *forest I, orchard type* and *longitude* are also present on this axis. The *orchard type* and the *latitude* are present on the axis 3, but axis 4 is purely geographical.

Table 8. Regression coefficients (CA) and canonical (CCA). The 3 higher indexes (absolute value) are indicated in bold

	Regression coefficients (CA)				canonical coefficients (CCA)			
	CA				CCA			
Axis	1	2	3	4	1	2	3	4
Descriptors								
Orchard type	0.04	0.53	**0.66**	0.08	-0.01	**0.84**	**0.68**	**-0.28**
Forest I	**0.26**	**0.78**	-0.05	-0.16	**0.29**	**0.76**	**-0.89**	0.05
Forest II	**0.32**	-0.06	0.11	**0.18**	**0.34**	0.05	0.52	0.28
X	-0.15	**-0.66**	**0.46**	**0.52**	-0.12	-0.65	0.55	**1.79**
Y	**-0.43**	**0.93**	**-0.47**	**-0.41**	**-0.49**	**0.93**	**-0.84**	**-1.12**

It is noteworthy that for the 17 associated species of the guild, the interpretation of axis 1 and 2 for the CA and the CCA (the most information worthy) remains the same. *Latitude* fixes axis 1 of the species distribution and the *orchard type* fixes the axis 2.

Following these observations, we can conclude that the axis for guild CA and CCA in 1992 has a very clear meaning with the same environmental descriptors. Axis 1 is fixed by the *latitude* (spatial descriptor) and by *forest II* (landscape descriptor). Analysis points out *forest I* and *orchard type* on the axis 2 (correlation coefficients in which the *orchard type* does not appear). *Orchard type* is also important on axis 3 and axis 4 is longitudinal.

The main difference between the results of 1992 and 1993 is the importance of the *landscape* descriptors in 1992, which is lost in 1993 (Fig. 7). The main environmental influence in 1993 is due to the geographical factor (*longitude*) and the *orchard type* factor. *Partial canonical analysis.*

We observed that movement direction (*emigration-immigration*) as well as orientation had no significant influence on the species matrix. After forward selection of the *landscape I* descriptor classes, the only class which can be taken into consideration in our model is the *forest* class. We obtained the same result with *landscape II* descriptor. On the level of analysis, it is interesting to put environmental descriptors into competition through a series of partial canonical analysis. The partial CCA are conducted with following descriptors: *orchard type, landscape I* (represented by the *forest*), *landscape II* (*forest*) and *geographical coordinates*.

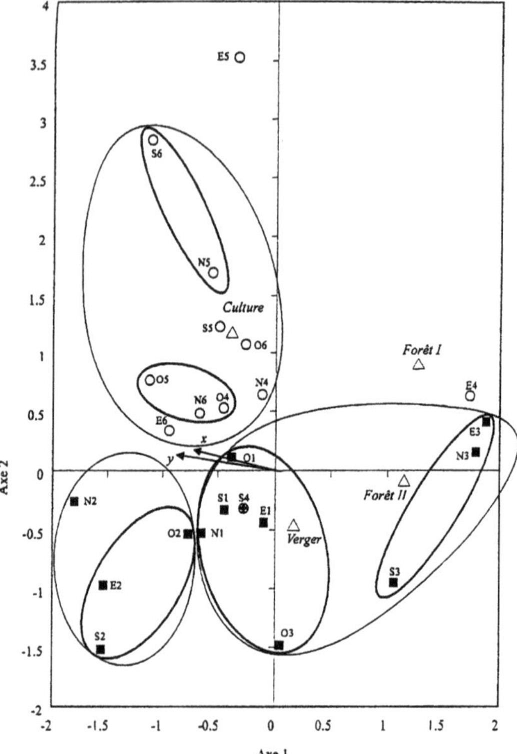

Figure 6. CCA diagram of the Malaise traps and of 4 environmental descriptors (*orchad type*, *landscape I*, *landscape II* and *geographical coordinates*). O : crops, ■ : orchards. 1, 2, 3, 4, 5, 6: orchards/crops. N, E, S, O: Malaise tents are set at the North, at the East, at the South and at the West, respectively. Nominal environmental descriptors classes are positioned as centroids of each sampling for every class (symbol: Δ). *Geographical coordinates* (x and y) are represented with an arrow which indicates direction of the greater variation. The shortest spanning tree (single linkage clustering) has been overlapped to define primary groups. Thick ellipsis: distance < 0.6; thin ellipsis: 0.6 ≤ distance < 0.7. Elements which are stressed with a cross do not belong to the group defined by the circle but are also set together.

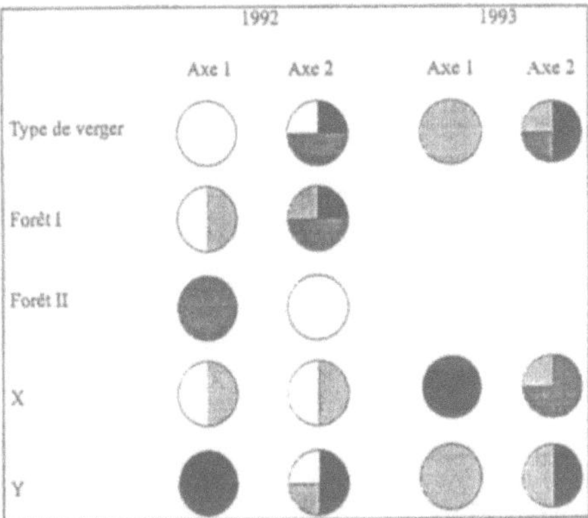

Figure 7. Summary of the value of the intraset correlation coefficients (CA and CCA) as well as regression coefficents (CA) and canonical (CCA) taken into consideration for each descriptor on the axis 1 and 2. Black = rank 1; dark grey = rank 2; pale grey = rank 3; white = not in the first 3 ranks. One fourth of the pie chart represents one rank obtained (4 coefficients calculated for each axis and each year).

The part of the variation commonly explained by the environmental descriptor and the space descriptor (b) is equal to 21.8 - 16.9 = 4.9% (= 13.7 - 8.8). Calculation of the unexplained part is: d = 100 - (16.9 - 8.8 - 4.9) = 69.4%. This is an important part and may be explained by environmental factors that have been eliminated, but were actually important elements in the explanation of variation in the guild matrix. It may also be explained by other factors not introduced, like the history of the orchard and its surrounding landscape, which could have an influence, but were not measured.

In this first series of analysis, we have extracted the variation that was explained by the *geographical coordinates* (8.8%), and tested this part by means of the Monte Carlo test (Table 9). The portion of variation of the guild species matrix explained by the *geographical coordinates* is significant (*p = 0.01*). If we look at Table 9, we can deduce that more or less 80% (16.9/21.8 x 100) of the environmental variation is due to local effects and is not spatially influenced. On another hand, two thirds (8.8/13.7 x 100) of the variation explained by the *geographical coordinates* is not dependent on the environmental descriptors. Less than one fourth (4.9/21.8 x 100) of the variation explained by the environmental descriptors can also be explained by the *geographical coordinates*. On this basis, we can deduce that the subjacent spatial structure common to the environmental factors and the guild is weak.

Table 9. Partial CCA I. Analysis is done on a 17 species matrix caught by means of 48 collectors, assembled two by two, on 24 Malaise traps set on the margin of 6 orchard/cultures. a: environmental variation (*landscape I*, *landscape II* and *orchard type*), c : spatial variation (*geographical coordinates*). *: significant values according to the Monte Carlo test with 99 permutations ($p \leq 0.01$)

	CCA first step	CCA second step
Total inertia = sum of unconstrained eigenvalues	2.532	2.532
Sum of canonical eigenvalues	0.551	0.348
Percentage of variation	21.8	13.7
	CCA third step	CCA fourth step
Total inertia	2.532	2.532
Sum of unconstrained eigenvalues[1]	2.185	1.981
Sum of canonical eigenvalues[1]	0.428	0.224
Percentage of variation	a = 16.9 *	c = 8.8

[1] after extraction of covariables

The second series of partial analysis separates the variation due to *landscape I* and *II* (taken together) and the variation due to the *orchard type* (Table 10). We completed this analysis with the Monte Carlo test in order to measure the statistical meaning of the influence of the environmental desriptors. This analysis shows that the *orchard type* significantly influences the matrix variation of the species belonging to ($p = 0.01$), as well as the *landscape* ($p = 0.01$).

Table 10. Partial CCA II . Analysis is done on a 17 species matrix caught in 48 collectors, assembled two by two, on 24 Malaise traps set on the margin of 6 orchard/culture interfaces. a_1 : *landscape* variation (*landscape I*, *landscape II*); a_3 : *orchard type* variation. * : significant values ($p \leq 0.01$)

	CCA first step	CCA second step
Total inertia	2.532	2.532
Sum of unconstrained eigenvalues[1]	2.185	2185
Sum of canonical eigenvalues[1]	0.267	0.162
Percentage of variation	10.5	6.4
	CCA third step	CCA fourth step
Total inertia	2.532	2.532
Sum of unconstrained eigenvalues[1]	2.023	1.918
Sum of canonical eigenvalues[1]	0.266	0.161
Percentage of variation	a_1 = 10.5 *	a_3 = 6.4 *

[1] after extraction of the covariables.

It is important to note that the part common to the *orchard type* and to the *landscape* is equal to 0. There is therefore no matrix guild that is common to the *orchard type* and to the *landscape*. If we interpret this result according to Borcard *et al.* (1992), this would imply that the landscape structure cannot explain changes of *orchard type* from one site to another and at the same time the guild variation The two environmental descriptors are completely independent. Partition of the total variation of the tortricid matri guild is summarised in Fig. 8.

The last series of partial canonical analysis are founded on the extraction and separation of respective influences of *landscape I* and *landscape II* (Table 11). *Landscape I* is responsible for 7.0% of the variation of the guild matrix and *landscape II* for only 2.4%. The Monte Carlo test indicates that the share due to *landscape I* is significant ($p = 0.001$), the share for *landscape II* has no influence ($p = 0.09$). *Landscape I* explains an important share of matrix variation (7.0%), this is not the same for *landscape II* which does not explain a significant share of variation (2.4%).

Table 11. Partial CCA III. Analysis is done on a 17 species matrix caught in 48 collectors, assembled two by two, on 24 Malaise traps set on the edge of 6 orchard/crop interfaces. a_{11} : *landscape I* variation ; a_{13} : *landscape II* variation. * : significant values ($p < 0.001$)

	CCA first step	CCA second step
Total inertia	2.532	2.532
Sum of unconstrained eigenvalues[1]	2.023	2.023
Sum of canonical eigenvalues[1]	0.205	0.09
Percentage of variation	8.1	3.6
	CCA third step	**CCA fourth step**
Total inertia	2.532	2.532
Sum of unconstrained eigenvalues[1]	2.023	1.918
Sum of canonical eigenvalues[1]	0.176	0.061
Percentage of variation	a_{11} = 7.0 *	a_{13} = 2.4

[1] after extraction of the covariables.

Compared with the results obtained with the whole tortricid community (results summarized in the Table 12, where the data of 1993 were also added), the guild is less sensitive to a an eventual biogeographical drift on a regional scale, this is shown by the *geographical coordinates*. Results can logically be interpreted in the following way: a guild finds a suitable environment that is, primarily the orchard (cultivated or not) all along the La Côte region and the species pool is identical. The difference in the trapping results is essentially due to more local factors, like the surroundings of the orchard. As for the whole population, which is essentially influenced by non apple tree associated species that depend on other environments, for example *landscape I* and *landscape II*, it is more sensitive to the biogeographical drift.

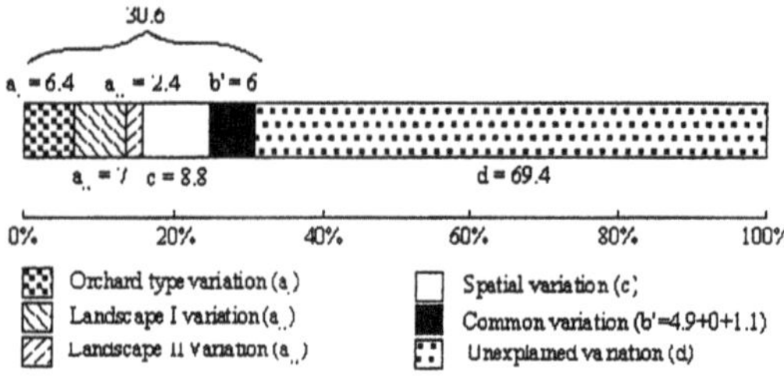

Figure 8. Summary of the partition of the total variation of the matrix guild. Results are drawn on the basis of partial CCA I, II, and III (Tab. 9, 10 and 11). a_{11}: *landscape I*, a_{13}: *landscape II*, a_3: *orchard type*, b': sum of common variation, c: spatial variation, d: unexplained variation.

Spatial components of the guild variation are slightly weaker for the whole of the settlement, but about the same size (8.8 and 11.9%). We can therefore say that the tortricid guild of apple tree associated species reacts to the geographical variation for the studied region in the same way as the whole of the tortricid population. Nevertheless, environmental variation (*landscape I*, *landscape II* and *orchard type*) is higher (16.9% vs. 13.7% for the whole population).

If we compare the results obtained with the whole population, the variation due to environmental descriptors is stronger because the *landscape* has a greater influence (10.5% for the guild and 7.1% for the population). Percentage explained by the *orchard type* is also stronger for the guild (6.4% respectively 5.9%). This means that tortricid guild activity on the interface is more sensitive to the general conditions in the orchard/crops and surrounding landscape.

Table 12. Summary of the explained variation percentage by the environmental descriptors on the population and guild matrixes in 1992 and 1993. *: significant *p*-value, non significant *p*-values are given, n.t.: non tested share

	Population				Guild			
	1992[1]		1993[2]		1992[1]		1993	
Environmental descriptors	%	p	%	p	%	p	%	p
Orchard type	5.9	a	6.7	*	6.4	*	2.9	0.09
Landscape I	4.8	*	2.4	*0.07*	7.0	*	non select.	
Landscape II	1.7	*0.3*	2.9	*0.01*	2.4	*0.1*	non select.	
Geographical coordinates	11.9	*	9.8	*	8.8	*	9.2	*
Common variation	6.9	n.t.	7.6	n.t.	6.0	n.t.	3.1	n.t.
Non explained variation	68.8	n.t.	70.6	n.t.	69.4	n.t.	84.8	n.t

[1]: *landscape I* and *landscape II* are represented by the *forest* class
[2]: *landscape I* and *landscape II* are represented by the *vineyard* class

To summarize, the share of variation that can be attributed to the environmental descriptors is more important for the apple tree associated guild than for the whole population. This difference is equally distributed between *landscape* descriptors and *orchard type*. On the other hand, the share that can be attributed to the *geographical coordinates* is more important in the case of the population.

Difference between the two trapping seasons is marked by the disappearance of the *landscape I* effect in 1993. There was a slight difference in trapping period between 1992 and 1993. In 1993, trapping began in June, while in 1992 it began one month later. Individuals caught in June have an influence on the proportion of individuals belonging to the associated species between the two types of orchards. If we compare these results with the ones obtained through the variance partition, we can deduce that by modifying the share of associated individuals in favor of the traditional orchards, we loose the influence of the *landscape I*. Consequently, *landscape* factors, particularly the orchard/crop adjacent environments have a greater influence on crops.

6.5 Concluding Remarks

Some common features of the results may be surprising. The amount of unexplained variation, for instance, is always fairly high. One cannot, however, discriminate between the potentially explainable variation and the real stochasticity in that unexplained variation. It may not be feasible to measure all the environmental variables (in the broad sense: biological interactions and external environmental factors) that are relevant in an ecological study. The amounts of variation involved in the main explained trends of the example data sets may seem proportionally low, but the underlying causes found to be significant can nevertheless be considered as important in the structuring of the tortricid community.

The more applied an ecological study is, the more the emphasis there is on the effects of particular environmental factors on ecological communities. Correspondingly, the statistical analysis should not just show the major variation in the species assemblages, but focus on the effects on the variables of prime interest (environmental factors). In our case, this was performed with CCA and partial CCA.

The analysis used in this study has shown that measured environmental descriptors act as explanatory factors on tortricid activity at the interface between the orchard and the surrounding environment. Mainly, the tortricids are distributed along a geographical gradient along the Lemanic Basin (large scale influence), but are also strongly influenced by landscape descriptors, particularly at a fine scale (adjacent ecosystems).

Acknowledgement

This study was financially supported by the Swiss National Foundation for the Scientific Research (Project 31-31549.91) and the Swiss Federal Research Station for Plant Production of Changins, Nyon. The author is grateful to Mrs. V. Michel, E. Röthlisberger, Mr. P.J. Charmillot, N. Alipaz, D. Pasquier and E. Jaccard for the very substantial help during the field and the laboratory work. Thanks are also expressed to Mrs. A. Jeanneret for critical reading of the manuscript and translation.

6.6 References

Altieri, M. & Schmidt, L., 1986. The dynamics of colonizing arthropod communities at the interface of abandoned, organic and commercial apple orchards and adjacent woodland habitats. *Agriculture Ecosystems and Environment* **16**: 29-43.

Balachowsky, A.S., 1966. *Entomologie appliquée à l'agriculture. Tome II, Lépidoptères.* Masson et Cie, Paris, 1057 pp.

Baudry, J., 1989. Interactions between agricultural and ecological systems at the landscape level. *Agriculture Ecosystems and Environment* **27**: 119-130.

Blommers, L.H., 1994. Integrated pest management in European apple orchards. *Annual Review of Entomology* **39**: 213-224.

Borcard, D., Legendre, P. & Drapeau, P., 1992. Partialling out the spatial component of ecological variation. *Ecology* **73**: 1045-1055.

Brunner, J.F.; Walker, J.T.S. & Suckling, D.M., 1988. Dispersal of the light brown apple moth, *Epiphyas postvittana*, in New Zealand apple orchards. Department of scientific and industrial research and New Zealand apple and pear marketing board, New Zealand, 30 pp.

Burel, F., 1992. Effect of landscape structure and dynamics on species diversity in hedgerow networks. *Landscape Ecology* **6**: 161-174.

Burel, F. & Baudry, J., 1995. Farming landscapes and insects. In: Glen; D., Greaves, M.P. & Anderson, H.M. (Eds.), *Ecology and integrated farming systems.* John Wiley and Sons, Bristol, pp. 203-220.

Chambon, J.P., 1986. *Les tordeuses nuisibles en arboriculture fruitière.* INRA, Paris, 118 pp.

Dennis, P. & Fry, G.L.A., 1992. Field margins: can they enhance natural enemy population densities and general arthropod diversity on farmland? *Agriculture Ecosystems and Environment* **40**: 95-115.

Dickler, E., 1991. Tortricid pests of pome and stone fruits, Eurasian species. In: Geest, L.P.S. & Evenhuis, H.H. (Eds.), *Tortricid pests, their biology, natural enemies and control.* Elsevier, Amsterdam, pp. 435-452.

Duelli, P., Studer, M. & Marchand, I., 1989. The influence of the surroundings on arthropod diversity in maize fields. *Acta Phytopathologica Entomologica Hungarica* **24**: 73-76.

Duelli, P., Studer, M., Marchand, I. & Jakob, S., 1990. Population movements of arthropods between natural and cultivated areas. *Biological Conservation* **54**: 193-207.

Fassotte, C., Chambon, J.P., Cocquempot, C.; Frerot B., Lettere M., Oger R. & Descoins, C., 1992. Détection de diverses espèces de Lépidoptères en vergers de pommiers (et alentours) par piégeage sexuel. *Mémoires de la Société royale belge d'Entomologie* **35**: 275-282.

Graaf Bentink, G.A. & Diakonoff, A., 1968. Die Nederlande Bladrollers (Tortricidae). Monografieën van de nederlandsche Entomologische Vereeniging, 3. Amsterdam, 201 pp.

Gulinck, H., 1986. Landscape ecological aspects of agroecosystems. *Agriculture Ecosystems and Environment* **16**: 79-86.

Hanneman, H.J., 1961. Kleinerschmetterlinge oder Microlepidoptera. I. Die Wickler (s. str.) Tortricinae. Die Tierwelt Deutschlands und der angrenzenden Meeresteile, 48. Teil. VEB Gustav Fisher, Jena, 233 pp.

Hope, A.C.A., 1968. A simplified Monte Carlo significance test procedure. *J. Roy. Stat. Soc. Ser. B* **30**: 582-598.

Jongman, R.H.G., Ter Braak, C.J.F. & Van Tongeren, O.F.R., 1987. *Data Analysis in Community and Landscape Ecology.* Pudoc, Wageningen, 299 pp.

Kuznetsov, V.I., 1989. 21. Tortricidae, (Olethreutidae, Cochylidae) - Tortricid moths. In: Medvedev, G.S. (Ed.), *Key to the insects of the European part of the USSR, Vol. IV Lepidoptera*, Part I, E.J. Brill, Leiden, pp. 279-956.

Legendre, L. & Legendre P., 1984. *Ecologie numérique. 1. Le traitement multiple des données écologiques.* 2e édition, Masson et Presses du l'Université du Québec, Paris et Québec, 260 pp.

Legendre, P. & Vaudor, A., 1991. *Le progiciel "R" - Analyse multidimensionnelle, analyse spatiale.* Département de Sciences Biologiques, Université de Montréal, Montréal, 144 pp.

Mani, E. & Wildbolz, T., 1977. The dispersal of male codling moths (*Laspeyresia pomonella* L.) in the upper Rhine Valley. *Journal of Applied Entomology* **83**: 161-168.

Scherrer, B., 1984. *Biostatistique*. Gaëtan Morin, éditeur, Canada, 850 pp.

Stinner, R.E., Barfield, C.S,; Stimac, J.L. & Dohse, L., 1983. Dispersal and movement of insect pests. *Annual Review of Entomology* **2**: 319-335.

Sziraki, G., 1984. Dispersion and movement activity of some important moth pests living on stone fruits. *Acta Phytopathologica Entomologica Hungarica* **19**: 51-64.

Ter Braak, C.J.F., 1986. Canonical correspondence analysis: a new eigenvector technique for multivariate direct gradient analysis. *Ecology* **67**, 1167-1179.

Ter Braak, C.J.F., 1987a. Unimodal models to relate species to environment. Doctoral thesis, Agricultural Mathematics Group-DLO, Wageningen, the Netherlands, 152 pp.

Ter Braak, C.J.F., 1990. Updates Notes: CANOCO Version 3.10. Agricultural Mathematics Group-DLO, Wageningen, the Netherlands. 35 pp.

Ter Braak, C.J.F., 1991. Update Notes: CANOCO Version 3.12. Agricultural Mathematics Group-DLO, Wageningen, the Netherlands. 11 pp.

Ter Braak, C.J.F., 1992. *CANOCO - a FORTRAN program for Canonical Community Ordination.* Microcomputer Power, Ithaca, New York, 95 pp.

Townes, H., 1972. A light-weight Malaise trap. *Entomological News* **83**: 239-247.

CHAPTER 7

LANDSCAPE CONNECTIVITY
Linking Fine-Scale Movements and Large-Scale Patterns of Distributions of Damselflies

PHIL D. TAYLOR
Atlantic Cooperative Wildlife Ecology Research Network.
Department of Biology, Acadia University, Wolfville, NS, Canada

7.1 Introduction

At least since Andrewartha and Birch (1954) ecologists have recognized that movement plays a crucial role in the dynamics of many populations. Movement is critical at an individual level in allowing animals to access heterogeneously distributed resources. At a population level, it is necessary for the establishment and re-establishment of local populations. Much of the development of our current conceptual framework of both metapopulation dynamics (Hanski, 1996) and the dynamics of other types of spatially-structured populations (Harrison, 1994) stems from this increased understanding of the importance of movement in animal population dynamics. At the same time, there has been an increasing recognition that scales of observation, experimentation and process are vital to understanding ecological principles and processes (Levin, 1992). As a result of that recognition, considerable effort has been expended conceptualizing and developing ideas about the process of movement and how it varies with scale. Unfortunately, empirical explorations of its importance to populations, and its relative importance at different spatial scales, have not kept pace with these theoretical studies.

Questions about the process of movement, and how movement relates to population dynamics, are embodied in the study of landscape connectivity. This paper is both an overview of the development of some of the ideas surrounding landscape connectivity, and of empirical work my students and I are undertaking. We are addressing questions about connectivity and landscapes, but we are also beginning to ask questions about the relative importance of connectivity in predicting patterns of distribution of organisms at a variety of spatial scales.

7.2 Landscape Connectivity

An essential part of the development of theory related to landscape connectivity was the notion that the *degree* to which local populations were joined, or connected, had an influence on the persistence of the metapopulation (*e.g.* Fahrig & Merriam, 1985; Lefkovitch & Fahrig, 1985). These ideas emerged from empirical work on the movements

B. Ekbom, M. Irwin and Y. Robert (eds.), Interchanges of Insects, 109-122

of small mammals in fencerows that showed that patches of habitat connected by linear elements improved the probability of regional population survival for some species (Henderson *et al.*, 1985).

From these early explorations came at least two important ideas. First, the work suggested that links between habitats - so-called corridors - were a potential conservation tool for aiding the persistence of spatially structured populations (Merriam, 1991). Second, and more generally, the work demonstrated, through a series of modelling and empirical surveys, that the extent to which patches of habitat were connected, and the 'quality' of those connections, positively influenced the probability of metapopulation persistence (Fahrig & Merriam, 1985; Henein & Merriam, 1990; Merriam, 1991; Fahrig & Merriam, 1994). Considerable controversy subsequently surrounded the first idea (*e.g.* Simberloff *et al.*, 1992) but, at least in part, such controversy rested on the false understanding that increasing the connectivity of a landscape was equivalent to the establishment of corridors (*e.g.* Forman, 1995). The second idea has gained more general acceptance in the current literature, even though it seems to have an even weaker base in empirical study. It is these latter ideas about landscape connectivity that I discuss here.

7.2.1 WHAT ARE THE ELEMENTS OF CONNECTIVITY?

Merriam (1984), and Henein and Merriam (1990) define connectivity as the extent to which an organism is able to move through the landscape to access resources vital to its survival. Taylor *et al.* (1993) generalized the concept to include it as a component of landscape structure (*sensu* Dunning *et al., 1992*). Most authors follow from this earlier work and include in definitions of connectivity two important elements: those related to the physical structure of the landscape and those related to behavior. Henein and Merriam (1990) recognized that ability to use resource patches was directly a function of both distance between patches (a measure of landscape structure) and the biology and behavior of the organism. With *et al.* (1997) define connectivity as a 'functional' link among habitats. They highlight the importance of the habitat elements and the dispersal capabilities of the organism. Schippers *et al.* (1996), and Hof and Flather (1996) consider connectivity to be a measure of the probability of individuals successfully immigrating to a new patch. Hof & Flather (1996) succinctly define connectivity mathematically using three parameters: the distance between patches and the nature of the matrix between patches, (both of which are measures of landscape structure) and the capability of dispersal in the species, a measure of behavior that interacts with structure.

In turn, the behavior of an organism at any given time is a function of the structure of the landscape; that is, animals will respond differently to different types of habitat elements within landscapes (Wiens *et al.*1997). Behavior will also interact with landscape structure at different spatial scales; these scales necessarily define the types of behavior and the types of interactions that we may observe (Holling, 1992). It is important to note that behavioral decisions at fine spatial or temporal scales may translate into differences in connectivity at larger-scales. In other words, the implications of a behavioral decision of an individual in a landscape at a fine scale may have consequences that extend far beyond that scale (Levin, 1992). Recently, some have recognized that there is room for fruitful interaction among the fields of landscape and behavioral ecology (*e.g.* Lima &

Zollner, 1996). Another way of considering this proposition is that there is an important need to consider how spatial and temporal scale influences measures of connectivity. Although scale is implicitly considered in the definitions of connectivity above, the idea that connectivity both varies with, and may mean different things at different spatial scales has only recently been explored (Keitt *et al.*, 1997).

Connectivity then, is an element of landscape structure; a function of the composition and configuration of the landscape, and a function of the relative ease by which an organism moves through elements in that landscape. It is a probabilistic measure that is determined by both fine-scale and large-scale patterns of behavior of the organism, that in turn are a response to fine and large-scale aspects of the physical structure of the landscape. Conceptually, it is a measure of the ease with which an organism moves through the landscape, taking into account that landscapes are heterogeneous mosaics of resources and non-resources. It incorporates both the amount of movement (the relative probability of an individual moving) and success of movement (the relative probability of an animal successfully moving between two points).

Implicitly or explicitly, connectivity usually refers to a population-level process. That is, it refers to the relative ease with which any two populations may exchange genetic material. Such a definition is clearly a function of 'connectivity' at a finer scale. However, it is not yet clear to what extent connectivity of landscapes at large scales is simply an additive function of connectivity at finer scales or whether the relationship between connectivity and scale is more complex (Keitt *et al.*, 1997).

7.2.2 HOW IS CONNECTIVITY MEASURED?

Connectivity can be inferred, modelled, or assessed empirically. As the distances that organisms are capable of moving increases, the spatial scale of observation increases, and the empirical assessment of the connectivity of landscapes becomes more difficult. As a consequence, most work has proceeded conceptually (Dunning *et al.*, 1992; Taylor *et al.*, 1993), using simulation modelling (Schumaker, 1996) null-model studies (With & Crist, 1996) or combinations of simulations and null models (Keitt *et al.*, 1997).

Empirically, connectivity has been inferred, but rarely explicitly measured. Hjerman & Ims (1996), for example, inferred connectivity for a European bush cricket as equivalent to the relative density of dispersing animals. Roland & Taylor (1997) inferred from meso- and large-scale surveys that connectivity differed among a group of four parasitoid species since the rate of parasitism by each species was influenced by forest structure (the relative proportion of the landscape that was forested) at different spatial scales.

Connectivity has been empirically assessed for even fewer animals, and usually at relatively small spatial and temporal scales (*e.g.* Wiens *et al.*, 1995). Rarely has the concept been explicitly studied at the meso-scales that are relevant to population processes (Kareiva & Wennergren, 1995; May, 1994). Exceptions include the work on small mammals described above (Merriam) and the work of Ilkka Hanski and his colleagues who have inferred, or directly measured connectivity in populations of the butterfly *M. cinxia* (*e.g.* Hanski *et al.*, 1994). Considerable additional work on colonization of patches has been undertaken by numerous researchers, but rarely with an explicit view to examining the process of movement through heterogeneous landscapes.

7.2.3 CONNECTIVITY AND SCALE

Movements that are normally considered trivial (*e.g.*, see Rankin & Burchsted, 1992 for insects) can lead to movement outside an organism's ecological neighbourhood (sensu Addicott, 1987; *e.g.*, Baars, 1979; Wegner & Merriam, 1990). For example, consider an organism such as a damselfly that breeds at streams, but forages in forest. If an individual attempts to fly to forest in a landscape largely devoid of forest, it may end up far from its 'home' stream at some neighbouring stream. An individual that makes such a movement has the potential to contribute to population dynamics at a larger spatial scale. To continue the example, if an inseminated female damselfly moves from one local population along a stream to an adjacent stream, it may colonize or recolonize that stream, establishing a new local population. This can happen even if the scale of the original movement was 'trivial' - that is, even if the animal was only attempting to locate forest for foraging. To more explicitly explore how these concepts of connectivity vary with and across temporal and spatial scales we need to recognize that there is enormous utility in learning how 'trivial', or fine-scale animal movement, is influenced by landscape structure. We also need many more empirical studies that explore the relationship between movement and landscape structure at multiple spatial and temporal scales.

7.3 Empirical Measures using Damselflies as Examples

7.3.1 OVERVIEW OF APPROACH AND METHODS

We have studied connectivity at three spatial scales in systems of Calopterygid damselflies that inhabit forested streams. *Calopteryx maculata* inhabits flowing streams as nymphs. Along these streams, adult males hold territories, and mate with females, and adult females oviposit on emergent vegetation (Waage, 1972). At the fine scale, over the course of a day, individual *C. maculata* move over distances that rarely take them more than 600 m from the edge of a stream. At this scale, we have measured the distribution of adults, and their ages, and have experimentally assessed how individuals from different types of landscapes move through different kinds of habitats. At a much larger scale (30 x 30 km) we have surveyed the distribution of animals in forest and agricultural landscapes. Using simple simulation models that exploit observed behaviors, or using inference from fine-scale studies, we have also begun to explore questions about the relative importance of fine-scale versus medium-scale processes in predicting the regional distributions of damselflies.

An overview of these studies is presented in this chapter. The results are preliminary, and should be viewed as a set of examples that demonstrate an overall approach, rather than the final word on landscape connectivity, or landscape connectivity in damselflies! In general, our approach has been to compare patterns or processes in landscapes that are predominantly forested (forest landscapes) to those same patterns and processes in agricultural landscapes that include open fields or pasture (mixed landscapes). *C. maculata* forages in forest in both types of landscapes, but in the mixed landscapes we have studied, forest patches are disjunct from the edges of streams by distances varying from 200-500 m.

Early in our studies, we observed some animals flying across the open pasture towards forest in these mixed landscapes. The resources necessary for damselfly survival (stream and forest) were distributed heterogeneously within these landscapes, and separated by inhospitable, or at least neutral, habitat (Taylor & Merriam, 1995). The observations suggested that the damselflies could persist in mixed landscapes by flying across open pasture linking their necessary resource patches together. These general observations gave rise to the series of surveys and experiments described below that aim to more clearly define connectivity for these damselflies at multiple spatial scales.

7.3.2 FINE-SCALE: PATTERNS OF DISTRIBUTION AROUND STREAMS

The initial observations of *C. maculata* flying across open fields to nearby forest suggested that connectivity might be an important component of the population dynamics of these organisms. We were first interested in comparing the distribution of animals around streams through mixed and forest landscapes. We wished to determine how far from streams *C. maculata* was found, how distance was related to age, and whether these relationships differed in mixed versus forest landscapes.

We initially conducted surveys of the distribution of *C. maculata* in the vicinity of the streams by doing a series of transects that extended perpendicular from the edges of streams in both forest and mixed landscapes for 600 m (for details see Taylor & Merriam, 1995). On average, animals were distributed farther from the streams in the mixed landscapes (284 ± 5 m; mean ± SE) than they were in the forest landscapes (189 ± 3 m). We rarely observed *C. maculata* in pasture habitats either on the surveys or in incidental observations. When we did observe damselflies in pasture habitat it was almost always individuals flying toward forest habitat or toward the stream - that is, they were transients within those habitats, rather than exploiting the habitats for food or shelter.

The results of the surveys also suggested that in forest landscapes, the length of the transects covered the bulk of the distribution of damselfly movements whereas in mixed landscapes, transects did not. Given an approximate separation of forest from stream in mixed landscapes of 200 m and adding the median distance *C. maculata* were observed from the edge of the stream in forest habitat (189 m) gives a conservative estimate of an expected median distance of *ca.* 389 m from the stream edge for mixed landscapes. This is a approximately 100 m greater than the observed value of 284 m suggesting that in mixed landscapes, our transects may not have encompassed the entire range of the distribution of adult *C. macaulata* away from the stream. Some adult *C. maculata* may move over distances exceeding 600 m from the stream edge to forage

During their teneral phase, newly emerged damselflies are known to move through landscapes and are often found far from streams (Corbet, 1980). During this maturation phase, which lasts between 7-11 days, the ovaries of female *C. maculata* will increase in size, and individual ovarioles change in shape and size (Johnson, 1973; Corbet, 1980). We therefore were able to assess how the age of individuals varied with distance by sampling females in both landscapes, and estimating age based on ovariolar state. On two days, we collected 16 females from each of a forest and mixed landscape, at different distances from the edge of the stream, (but not at the stream). Sampling effort was balanced across landscape and distance.

Details of dissections and ovariolar assessment are as presented in Taylor & Merriam (1996). We assessed ovariolar state using four ordinal categories and then tested the effects of ovarian state and type of landscape on the distance an animal was found from the edge of the stream using ANOVA. There was a significant effect of ovariolar state on distance from edge of stream (p(F) = 0.036). Younger females (with undeveloped ovaries) tend to be found farther from the edge of the stream than older females. The type of landscapes also had an effect on distance (p(F) < 0.001); *C. maculata* in the mixed landscape were found farther from the stream than in the forest landscape, a result consistent with the previous analysis. In both types of landscapes, at least some older individuals were still found at the *maximum* distances we surveyed from the edge of the stream.

In summary, *C. maculata* are found at distances up to 600 m from the stream in both forested and mixed landscapes, but these are likely conservative estimates of the maximum distances that animals can be found. As females mature, they are more likely to be found closer to the edges of the streams. The importance of these simple findings is several-fold. In the landscapes where we surveyed, approximate inter-stream distances ranged between 0.5 and 2 km. Therefore, any increase in distances that an individual moves as a consequence in of the structure of the landscapes may imply shorter inter-local populations distances than implied by simple geographic distance. Secondly, older, sexually mature individuals can and do move up to the maximum distance we observed (600 m) from the edges of the stream.

7.3.3 MEDIUM-SCALE: MEASURING CONNECTIVITY IN CALOPTERYX MACULATA

The observed fine-scale distributions in the mixed landscape strongly suggested that individual damselflies were moving easily between resource patches. In turn, this raised the interesting possibility that connectivity might be enhanced in these moderately fragmented landscapes. Enhanced connectivity would imply that in mixed landscapes, animals from different local populations (groups of damselflies found along individual streams) would interact (exchange genetic material) more frequently than in forest landscapes. What was important to know was whether, in this system, damselflies were impeded, aided, or not affected, by travelling through pasture elements in the mixed landscape. To answer the question, we designed and performed a manipulative field experiment. The field experiment was used to assess the relative ability of the organisms to move through two different elements in the landscape: forest and pasture, and to assess whether that ability differed as a function of the sex of the animal, and its natal landscape.

The basic design of the experiment was as follows. In each of two replicates, 20 male and 20 female *C. maculata* were collected from the edge of the stream in a forest and an adjacent mixed landscape for a total of 80 animals. Each animal was marked by writing a unique alphanumeric on its wings using thinned typewriter correcting fluid. The animals were placed in 1 m^3 cages (according to the design outlined below) to acclimatize for 1 h, then released. The main treatment was to displace animals from the edge of the stream (where mating and oviposition take place) by approximately 300 m, releasing them, and testing their ability to move through either a forest or a pasture habitat. The expected number to be re-observed in each landscape was provided by releasing half of the animals at the edge of the stream. Both natal and release landscape were controlled for by translocating half the individuals between landscapes. The design is shown graphically in Figure 1. The objective is not to determine the homing abilities of the animals, but to assess their ability to move through the different elements of the landscape as a function of their landscape of origin and their sex.

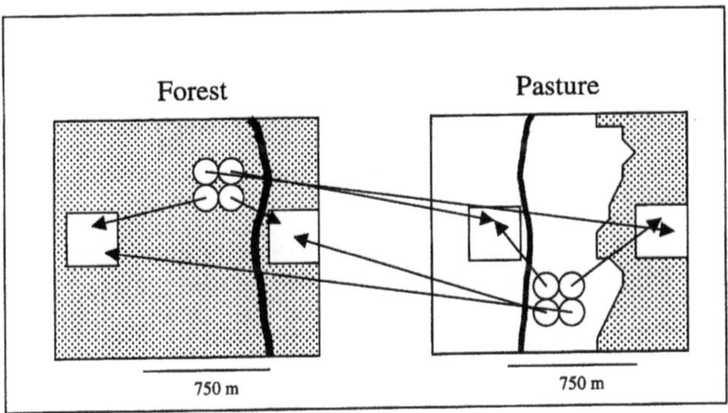

Figure 1. Schematic diagram of experimental manipulation to test the effect of source and release landscapes on the ability of damselflies to move through pasture and forest elements. Stippled areas are forested and open areas are pasture. The dark line represents the stream. Each circle represents 10 male and 10 female damselflies. Squares represent the cages that were situated at the streams, or 300 m distant from the stream. Arrows indicate how damselflies were moved within and between landscapes.

The results of the experiment were inconclusive. In brief, more than the expected number of displaced animals was re-observed in the mixed landscape than in the forest landscape but the result was only marginally significantly different from what would be expected by chance (logistic regression; Treatment x landscape term; $p(\chi^2) = 0.116$). Subsequently, we have repeated these experiments for both *C. maculata* and a congeneric species, *Calopteryx aequabilis* (Pither & Taylor, 1998) and have demonstrated a significant effect of the release landscape on the re-observation rate of *C. maculata*; pasture facilitates movement for *C. maculata*, and forest landscape impedes movement.

These findings raise additional questions about whether extensive flights across pasture are simply an extension of existing behavior, or are new behaviors that arise as a function of the new type of landscape. In particular, we were further interested in the role of fine-scale behavioral decisions for the measurement of connectivity, and the impact that these decisions have on landscape connectivity. The experiments to elucidate these decisions have been conducted in a similar fashion to the translocation experiments described above. Three types of landscapes are examined: those that are completely forested, those where forest is disjunct from the edge of the stream as in the experiments above and landscapes with little, or no forest. In these experiments we tested how release landscape, natal landscape, and sex influence the probability that an animal leaves the stream, and the pathways they take. We have experimentally translocated, released and tracked over 100 damselflies of two species (*C. maculata* and *C. aequabilis*) in each of 15 different landscapes, of the three types.

The results indicate that, for both species, the probability of animals leaving the stream is highest in the forest and mixed landscapes and lowest (near zero) in the non-forested landscapes (Jonsen & Taylor, 1999). Coupled with the ease of movement studies, these results suggest then that the connectivity of moderately fragmented landscapes is higher than both non-fragmented landscapes and completely fragmented landscapes. From a population point-of-view, this implies a metapopulation-like population structure in forest and completely fragmented landscapes (where local populations may be sufficiently isolated to have independent dynamics) but a patchy population structure in moderately fragmented landscapes.

7.3.4 LARGE-SCALE: REGIONAL PATTERNS OF DISTRIBUTION OF CALOPTERYX MACULATA

The fine and meso-scale studies presented above allowed us to assess how individuals react to landscape structure, but gave no insight into how, at larger spatial scales, local populations were distributed with respect to local, and meso-scale features of the landscapes. Consequently, we have undertaken large-scale surveys of the distribution of *C. maculata* to allow us to assess the relative importance of local and medium-scale habitat characteristics on the large-scale patterns of distribution of this forest damselfly. In addition, we are in the process of using these large-scale patterns of distributions to begin to explore how measures of connectivity might be used to predict where (in which kinds of landscapes) and when (under what kinds of climatic or environmental conditions) meso-scale features of the landscape such as proximity of forest, influence larger-scale patterns of distribution.

First, we surveyed the regional distribution of *C. maculata* within two 30 x 30 km landscapes in southern Ontario, Canada (northern landscape, southern landscape). We first marked all road-stream or road-drainage ditch intersections on topographic maps and then, over a one-week period at the peak of the flight season, each site was visited during the middle of the day (1000-1500) when damselflies are most active (Forsyth & Montgomerie, 1987). At each point we recorded the number of male and female *C. maculata* present, the presence or absence of flow in the stream, and the stream width. Eighty-five points were surveyed in the northern area and 147 points were surveyed in the southern area. We analyse these data as part of the following section.

7.3.5 LINKING SCALES: SIMULATION MODELS OF LANDSCAPE CONNECTIVITY FOR CALOPTERYX MACULATA

The empirical studies at three separate spatial scales that I described above, generate a number of interesting questions about landscape connectivity. In my opinion, the most important are questions about whether knowledge of behavior at either fine or medium scales can be used to predict patterns of distribution of *C. maculata* at larger spatial scales. That is, are the patterns of distribution observed at the large scale simply a function of available habitat (the local characteristics of the streams); are they a function of the particular pattern of resources available on the landscape (the relative size and positions of forest patches). Further, when information about the locations of forests is combined with knowledge of the fine- and meso-scale behaviors of the individuals, are our predictions of large-scale distributions improved?

To explore these questions, we developed a simple, individual-based simulation model of damselfly movement that was parameterized and tested using the empirical surveys and experiments outlined above. We then ran the simulation model using the real patterns of forest patches, open landscapes and streams from our large-scale empirical surveys and explored the relationship between the large-scale empirical survey results and local stream characteristics, medium-scale patterns of forest resources, and measures of connectivity estimated from our simulation models.

First, a LANDSAT image of the southern and northern landscapes was classified into three classes: forest, water and non-forest. Forest patches were then classified by size into three groups: patches >100 m^2; patches >1 km^2 and patches >10 km^2. In a geographic information system (GIS), we then created three maps showing the distance to the nearest forest patch from every point in the landscape. Similarly, all streams in the region were digitized from 1:50000 National Topographic Series maps and maps of stream presence, and distance to the nearest stream were constructed using GIS. Thus, for every point on the landscape we had knowledge of the presence or absence of, and distance to each of the two focal resource patches for *C. maculata*.

The structure of the simulation model was made deliberately simple since we were not interested in recreating the precise dynamics of damselfly movement, but rather, in obtaining a relatively simple way of predicting how movement influences distribution across scales. All damselflies were considered to be females, and each lived for 30 days. Each had an 8-day period as a teneral during which time it continually searched for forest, or if in forest, moved randomly. On each day from day 9 through day 30, individuals alternated between moving towards streams (for oviposition) and moving towards patches of forest (for foraging). After reaching a stream, individuals moved up or down the stream in a random direction for the remainder of a 'day'.

To run this model, we estimated three parameters from empirical data. The three parameters covered the range of movements described above: **a)** the mean distance a damselfly travelled from the edge of the stream into forest, (estimated from the fine-scale surveys); **b)** the maximum amount of pasture across which a damselfly could detect forest (estimated from fine-scale surveys and anecdotal observations) and **c)** the distance a damselfly moved along a stream when at the stream (estimated from unpublished marking studies).

The validity of the model was first tested by comparing the simulated distribution of damselflies in forest and forest-pasture sites with local-scale measures of density obtained from the 600 m transects perpendicular to streams described in the fine-scale surveys. The regional distribution of *C. maculata* was then simulated by populating each of the two simplified forest/stream landscapes presented above, with 1 individual per 50 m of stream and running the models for 30 simulated days. Measures of connectivity for a site along a given stream were computed by calculating the density of visits by all damselflies at that point in the landscape. The rationale was that the number of visits damselflies made to a point on the stream was a measure of how easily they could access that point.

7.3.6 LINKING SCALES: TESTING PREDICTIONS OF THE SIMULATION MODELS

Using the large-scale empirical data, we first tested the effects of region (northern or southern landscape) and stream flow on the presence or absence of *C. maculata* at each point by fitting a logistic regression model in Splus (StatSci corporation). The effects of both region and stream flow (habitat composition) were large and significant ($p(\chi^2) < 0.0001$). *C. maculata* are found more frequently north of the Ottawa river than south, and virtually all records of *C. maculata* were from flowing streams. Lack of stream flow partly explains why fewer streams in the south had *C. maculata* populations. Subsequent analyses were conducted on the subset of the data where streams flowed, and separately for the two landscapes. Proximity of medium (>1 km^2) and large (>10 km^2) forest patches increased the probability that *C. maculata* would be present at a stream in the southern landscape ($p(\chi^2) = 0.005$; $p(\chi^2) = 0.003$) but not in the northern landscape. *C. maculata* was present at 53/65 (82%) of survey points where medium-sized patches of forest were closer than 750 m. The was no significant additional contribution of the measure of landscape connectivity to the models in either southern or northern landscapes.

Two landscape-scale components were found to be overwhelmingly important in predicting the presence of *C. maculata* at streams: stream flow and proximity of medium-sized forest patches. There were additional large-scale differences between the northern and southern landscapes with respect to the presence of damselflies, but these differences did not change the influence of either stream flow or proximity of forest patches.

The additional predictive power of the simulated damselfly densities (landscape connectivity) was nil; we obtained our best predictions of the regional distribution of damselfly based only on our knowledge of local-scale response to landscape structure. The reason for the lack of any cross-scale effect may be quite simple. Proximity of forest and stream flow had such important influences on the presence of damselflies, that there was little additional variation available to be explained. Virtually all observations of *C. maculata* were at flowing streams, and few were more than 750 m from medium-sized patches of forest.

7.4 Landscape Connectivity for Damselflies

Collectively, our results above show that damselflies respond to medium-scale features of the landscape (the relative amounts of forest and open land) and suggest that landscape structure may have a strong influence on landscape connectivity for damselflies. Damselflies inhabiting streams through landscapes with a mixture of forest and pasture, are more likely to move off the stream, and then move more easily through pasture elements relative to their patterns of movement in forest. As a consequence, we expect that damselflies within mixed landscapes will have an expanded ecological neighbourhood relative to populations within completely forested or completely open landscapes. The important point is that the damselflies seem to be responding at a fine-scale (individual foraging behaviors) to a medium-scale feature of the landscape (but a feature within their perceptual range; see Zollner & Lima, 1997), and that because of the relative ease with which individuals can move through different types of habitat, these responses *may* influence population structure at even larger spatial scales.

However, using very simple simulation models we could not show that a measure connectivity improved our ability to predict the distribution of damselflies at a larger spatial scale. That is, the probability of a damselfly occupying a stream, in single year, in the landscapes we studied, was completely dependent on the local habitat features and the proximity of nearby forest. Because of the simplicity of these early models, we should not place too much weight on this negative evidence.

7.5 What Important Questions Remain?

One important area of further study is to examine how responses to landscape structure vary across taxa. For example, other studies we are undertaking focus on asking similar questions as outlined above, but using amphibians as models. Amphibian populations are similar to the damselfly system in that important resource requirements include both terrestrial and aquatic habitats. The rationale is to look at how ecological processes such as movement differ across the two very different groups of taxa. The results from a range of such studies will begin to address important questions about which kinds of species are most susceptible to anthropogenic changes in landscape structure.

A second important area, that we have only begun to address empirically, is how connectivity varies with spatial scale. The simulation models above suggest that it is possible to predict large-scale patterns of distribution based on knowledge of both fine and medium-scale features of the landscape, but it is uncertain when and where actual behavioral measures of connectivity might matter. Theoretical work in this area suggests that there might be important scale-influenced shifts in landscape connectivity that are a function of the interaction between patch placement and dispersal distances. Other recent work suggests that most animals respond only to loss of habitat (essentially a fine-scale feature from the point of view of an individual) and that elaborate measures of connectivity may be irrelevant (Andrén, 1994; Bender *et al.*, 1998). Whether these predictions hold out is an open question, but it is clear that current definitions of landscape connectivity do not deal explicitly with how spatial scale influences connectivity, and that important gains are to be made in empirical studies of these phenomena.

A third important area of increasing interest is to determine when connectivity is important to population persistence. Our research has shown us that Calopterygid damselflies are able to persist within moderately fragmented landscapes by linking together spatially separate resource patches. However, our initial results suggest that in highly fragmented landscapes, damselflies must switch to new resource patches for foraging, as sufficient forest no longer exists close to the streams along which they must mate, oviposit and develop. Are there critical amounts and distributions of resources within landscapes below which persistence begins to decline? Andrén (1994) suggested such a threshold effect may exist for birds and small mammals, but it remains to be discovered how such concepts can be more generally applied to other important landscape ecological processes.

Acknowledgements

Portions of this work have been supported by the following organizations and agencies: Wildlife Habitat Canada, Carleton University Graduate Studies, Natural Sciences and Engineering Research Council (Grants to H.G. Merriam and P.D. Taylor), the University of Alberta, the Atlantic Cooperative Wildlife Ecology Research Network, and the Acadia University Faculty Association. Significant contributions to the studies were made by Nicole Nadorozny, Jason Pither and Ian Jonsen. Numerous landowners in Nova Scotia, Ontario and Quebec have generously allowed us access to their properties to conduct our work. I thank John Wegner for comments on the manuscript.

7.6 References

Addicott, J.F., Aho, J.M., Antolin, M.F., Padilla, D.K., Richardson, J.S. & Soluk, D.A., 1987. Ecological neighbourhoods: scaling environmental patterns. *Oikos* **49**:340-346.

Andrén, H., 1994. Effects of habitat fragmentation on birds and mammals in landscapes with different proportions of suitable habitat: a review. *Oikos* **71**:355-366.

Andrewartha, H.G. & Birch, L.C., 1954. *The distribution and abundance of animals.* University of Chicago Press, Chicago.

Baars, M.A., 1979. Patterns of movement of radioactive carabid beetles. *Oecologia* **44**:125-140.

Bender, D.J., Contreras, T.A. & Fahrig, L., 1998. Habitat loss and population decline: a meta-analysis of the patch size effect. *Ecology* **79**:517-533.

Corbet, P.S., 1980. Biology of Odonata. *Annual Review of Entomology* **25**:189-217.

Dunning, J.B.; Danielson, J.B & Pulliam, H.R., 1992. Ecological processes that affect populations in complex landscapes. *Oikos* **65**:169-175.

Fahrig, L. & Merriam, G., 1985. Habitat patch connectivity and population survival. *Ecology* **66**:1762-1768.

Fahrig, L. & Merriam, G., 1994. Conservation of fragmented populations. *Conservation Biology* **8**:50-59.

Forman, R.T.T., 1995. *Land Mosaics: the ecology of landscape and regions.* Cambridge University Press, Cambridge, 632 pp.

Forsyth, A. & Montgomerie, R.D., 1987. Alternative reproductive tactics in the territorial damselfly Calopteryx maculata: sneaking by older males. *Behavioral Ecology and Sociobiology* **21**:73-81.

Hanski, I., 1996. Metapopulation ecology. In: Rhodes, O.E., Chesser, R.K. & Smith, M.H. (Eds.), *Population Dynamics in Ecological Space and Time.* University of Chicago Press, Chicago, 388 pp.

Hanski, I.; Kuussaari, M. & Neiminen, M, 1994. Metapopulation structure and migration in the butterfly *Melitaea cinxia*. *Ecology* **75**:747-762.

Harrison, S., 1994. Metapopulations and conservation. In: Edwards; P.J., May, R.M. & Webb, N.R. (Eds.), *Large-scale ecology and conservation biology.* Blackwell, Oxford.

Henderson, M.T., Merriam, G. & Wegner, J., 1985. Patchy environments and species survival: Chipmunks in an agricultural mosaic. *Biological Conservation* **31**:95-105.

Henein, K. & Merriam, G., 1990. The elements of connectivity where corridor quality is variable. *Landscape Ecology* **4**:157-170.

Hjerman, D.O. & Ims, R.A., 1996. Landscape Ecology of the Wart Biter *Decticus verrucivorus* in a patchy landscape. *Journal of Animal Ecology* **65**:768-780.

Hof, J. & Flather, C.H., 1996. Accounting for connectivity and spatial correlation in the optimal placement of wildlife habitat. *Ecological Modelling* **88**:143-155.

Holling, C.S., 1992. Cross-scale morphology, geometry, and dynamics of ecosystems. *Ecological Monographs* **62**:447-502.

Johnson, C., 1973. Ovarian development and age recognition in the damselfly, *Argia moesta* (Hagen, 1961) (Zygoptera: Coenagrionidae). *Odonatologica* **2**:69-81.

Jonsen, I.D. & Taylor, P.D., 1999. Fine-scale movement behaviors of Calopterygid Damselfliesare influenced by landscape structure: An experimental manipulation *Oikos*. (in press).

Kareiva, P. & Wennergren, U., 1995. Connecting landscape patterns to ecosystem and population processes. *Nature* **373**:299-302.

Keitt, T.H., Urban, D.L. & Milne, B.T., 1997. Detecting critical scales in fragmented landscapes. *Conservation Ecology [online]* **1(1)**: 4. URL: http://www.consecol.org/vol1/iss1/art4

Lefkovitch, L.P. & Fahrig, L., 1985. Spatial characteristics of habitat patches and population survival. *Ecological Modelling* **30**:297-308.

Levin, S.A., 1992. The problem of pattern and scale in ecology. *Ecology* **73**:1943-1967.

Lima, S.L & Zollner, P.A., 1996. Towards a behavioral ecology of ecological landscapes. *Trends in Ecology and Evolution* **11**:3-6.

May, R.M., 1994. The effects of spatial scale on ecological questions and answers. In: Edwards; P.J., May, R.M. & Webb, N.R. (Eds.), *Large-scale ecology and conservation biology*. Blackwell, Oxford.

Merriam, H.G., 1984. Connectivity: a fundamental ecological characteristic of landscape pattern. In: Brandt, J. & Agger, P., (Eds.), *Proceedings of the first international seminar on methodology in landscape ecological research and planning - Theme 1*. International Association for Landscape Ecology, Roskilde University, Roskilde.

Merriam, G., 1991. Corridors and connectivity: animal populations in heterogeneous environments. In: Saunders, D.A. & Hobbs, R.J. (Eds.), *Nature Conservation 2: The role of corridors*. Surrey Beatty & Sons, Chipping Norton.

Pither, J. & Taylor, P.D., 1999. An experimental assessment of landscape connectivity. *Oikos* **83**:166-174.

Rankin, M.A. & Burchsted, J.C.A, 1992. The cost of migration in insects. *Annual Review of Entomology* **37**:533-559.

Roland, J. & Taylor, P.D., 1997. Insect parasitoid species respond to forest structure at different spatial scales. *Nature* **386**:710-713.

Schippers, P., Verboom, J., Knaapen, J.P. & van Apeldoorn, R.C., 1996. Dispersal and habitat connectivity in complex heterogeneous landscapes: an analysis with a GIS-based random walk model. *Ecography* **19**:97-106.

Schumaker, N.H., 1996. Using landscape indices to predict habitat connectivity. *Ecology* **77**:1210-1225.

Simberloff, D., Farr, J.A., Cox, J & Mehlman, D.W., 1992. Movement corridors: Conservation bargains of poor investments? *Conservation Biology* **6**:493-504.

Taylor, P.D., Fahrig, L., Henein, K. & Merriam, G., 1993. Connectivity is a vital element of landscape structure. *Oikos* **68**:571-573.

Taylor, P.D. & Merriam, G., 1995. Wing morphology of a forest damselfly is related to landscape structure.

Oikos **73**:43-48.

Taylor, P.D., 1996. Habitat fragmentation and parasitism of a forest damselfly. *Landscape Ecology* **11**:181-189.

Waage, J.K., 1972. Longevity and mobility of adult *Calopteryx maculata* (Beauvois, 1805) (Zygoptera: Calopterygidae). *Odonatologica* **1**:155-162.

Wegner, J.F. & Merriam, G., 1990. Use of spatial elements in a farmlands mosaic by a woodland rodent. *Biological Conservation* **54**:263-276.

Wiens, J.A., Crist, T.O., With, K.A. & Milne, B.T., 1995. Fractal patterns of insect movement in microlandscape mosaics. *Ecology* **76**:663-666.

Wiens, J.A., Schooley, R.L. & Weeks, R.D. Jr., 1997. Patchy Landscapes and animals movements: do beetles percolate? *Oikos* **78**:257-264.

With, K.A. & Crist, T.O, 1996. Translating across scales: simulating species distributions as the aggregate response of individuals to heterogeneity. *Ecological modelling* **93**:125-137.

With, K.A., Gardner, R.H. & Turner, M.G., 1997. Landscape connectivity and population distributions in heterogeneous environments. *Oikos* **78**:151-169.

Zollner, P.A. & S.L. Lima., 1997. Landscape-level perceptual abilities in white-footed mice: perceptual range and the detection of forested habitat. *Oikos* **80**: 51-60

CHAPTER 8

EFFECT OF HABITAT HETEROGENEITY ON THE DIVERSITY AND DENSITY OF POLLINATING INSECTS

JÓZEF BANASZAK
Pedagogical University, Department of Biology and Environment Protection, Bydgoszcz, Poland

8.1 Introduction

This chapter presents some results of studies conducted by the author and his colleagues in the field of bee ecology at the landscape level, concerning natural bee resources in various ecosystems and rules governing the distribution of Apoidea. The studies have been carried out in natural and agricultural landscapes, mainly in the area of Poland and Rumania. The diversity and density of Apoidea have been investigated also in other parts of Europe: Bulgaria, Germany and France. The latest, unpublished data were collected by the author in Greece (Lefkada). All the studies aimed at determining an appropriate strategy for the conservation of pollinating insects and thus creating opportunities for maintaining the present variety and numbers of these important insects, if not actually increasing them. Bees are considered here in the broad sense, *i.e.* as the whole family Apoidea, which comprises both solitary and social wild bees, and the honey bee (*Apis mellifera* L.).

The first period of intensive development of European apidology was the nineteenth century, when lists of bee species inhabiting smaller or larger areas were compiled and taxonomic studies were carried out. In some countries, for example in Poland, these directions of research predominated until the 2nd World War. Today, studies on the taxonomy, biology and social behavior of Apoidea are the most intensively developing branches of bee science.

The 1950's witnessed a rising interest in pollinators of crops related to an increased demand for fodder plant seeds. However, these studies were hindered by two major obstacles. Firstly, the applied methods of bee density assessment did not produce objective results as a rule. Secondly, bee density on a crop does not give a true picture of bee numbers but is an accumulated index of the resources of the landscape surrounding the plantation.

With the continuing need for the pollinating activity of bees in farming, and the alarming opinions of ecologists about the state of the environment (Goldsmith & Hildyard, 1988), there is a need to evaluate the resources of pollinators. The studies should cover whole landscapes, including refuge areas of Apoidea at different degrees of transformation, and crop plantations in various sites. This direction of research would facilitate: (1) assessment of the present fauna; (2) prediction of changes in the fauna, their rate and directions;

B. Ekbom, M. Irwin and Y. Robert (eds.), Interchanges of Insects, 123-140

(3) appropriate regionalization of seed plantations; (4) prognosis of seed yield; etc (Banaszak, 1989). The quantitative analysis of whole landscapes is one of the latest trends in pollination ecology, and literature data in this field is still scarce.

8.2 Pollinating Insects and Current Surveys of their Resources

There is a common belief that pollinating insects constitute an important factor in increasing the yield of many crops. For fruit trees and shrubs, this function is successfully fulfilled by the honey bee. However, to pollinate some other crops, particularly fodder plants, participation of wild bees is necessary. An evaluation of the acreage of crop plantations, mean density of bees and their participation in pollination (Banaszak & Cierzniak, 1995) showed that annual yields of five crops in Poland in early 1990's as a result of pollination by Apoidea were as follows:

* lucerne seeds: 790.1 tons ($1.9 million);
* red clover seeds: 8200.0 tons ($21.5 million), see Fig. 1;
* oilseed rape: 53,928.0 tons ($5.5 million);
* buckwheat: 10,411.0 tons ($8.7 million);
* apples: 979,890.0 tons ($18.7-303.3 million, depending on their quality), see Fig. 2.

The total economic effect of pollination of those five crops amounts to $59.2-343.8 million. This includes $40.6-311.4 million contributed by the honey bee and $18.5-32.4 million contributed by wild bees.

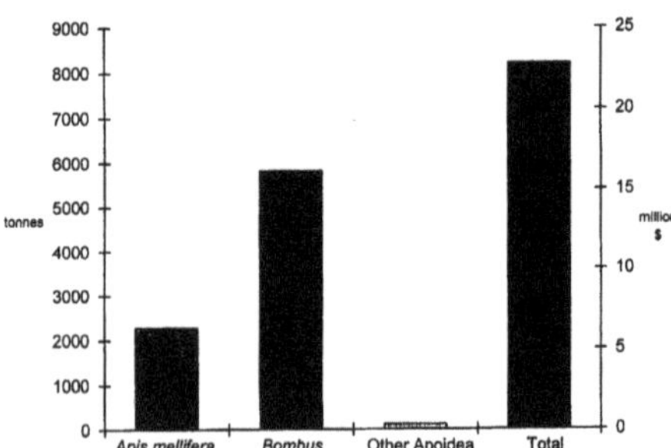

Figure 1. Yields of red clover seeds and the economic effect resulting from pollination by the honey bee and wild bees in Poland.

A 15-year study conducted in the Wielkopolska-Kujawy Lowland (mid-west Poland), which is a typically agricultural region, showed that despite a strong human pressure, the potential capacity for plant pollination by bees is considerable there due to a high degree of preservation of refuge habitats (Banaszak 1983, 1987). This area of about 20,000 km^2 was found to be inhabited by almost 260 species of Apoidea (Banaszak, 1983, 1987). It should be stressed that a similar number of bee species was recorded in this area 50 years earlier. In 1970's and 1980's, the diversity and density of Apoidea were studied in the same region in 6 sites located in natural and agricultural landscapes. After 10 years, higher values of both density and diversity were recorded in the majority of these sites (Banaszak, 1997). This allows for optimism and suggests that there exist some factors which compensate for the harmful impact of human activity on the fauna. (This topic will be discussed in detail in the next section).

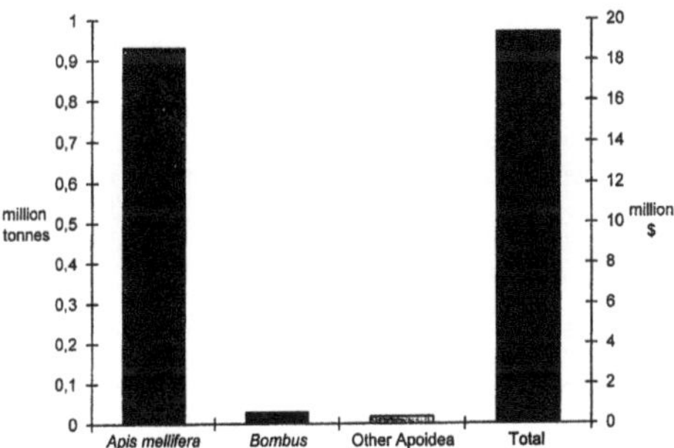

Figure 2. Yields of apples and the minimum economic effect resulting from pollination by the honey bee and wild bees in Poland.

Information on contemporary bee diversity is included in faunistic works which have been published for over a century both in the New World and in the Old World. Bee species inhabiting Eastern Europe are listed in Osytshnjuk*et al.* (1978), while those found in West Europe are described in Rasmont *et al.* (1995). Examples of assessment of local faunas are checklists of Apoidea of Finland - 273 species (Vikberg, 1986), Sweden - 278 (Janzon *et al.*,1991), Lithuania - 322 (Monsevicius, 1995), Poland - 454 (Banaszak, 1991), West Germany - 517 (Westrich, 1984), Czechoslovakia - 699 (Kocourek, 1989), Italy - 944 (Pagliano, 1995) and Spain - 976 (Ceballos, 1956). An example of a synthetic analysis of the bee fauna of Poland is a collective work edited by Banaszak (1992b). A comprehensive assessment of the contemporary fauna of Apoidea in Europe and an evaluation of its changes are given in Banaszak (1995).

For seed production, however, the density of pollinators in farming areas is of paramount importance. Recent research on the density of Apoidea on various crops in Poland shows that it is much below the optimum level. This concerns mostly solitary bees whose role in plant pollination is complementary with respect to the honey bee (Table 1). Because of the growing need for pollination of seed plantations, investigations into the density of Apoidea were started in the 1950's. However, no earlier than in late 1970's, the investigations started to include semi-natural and natural habitats, such as hedgerows, roadsides, woodlands and grasslands. The most popular method of bee density assessment is the line transect method (Banaszak, 1980). The mean density of wild Apoidea in natural habitats of the Wielkopolska National Park (mid-west Poland) was 256 individuals/ha. A similar density of wild bees was recorded in refuge environments of northern Germany. In crop plantations the density of Apoidea is several times higher. In south Europe bee densities also tend to be higher. In the Hungarian steppe, Tanacs (1982) recorded 760-1110 individuals/ha. Particularly high densities of Apoidea, which exceeded 1,000 individuals/ha, were observed in natural and semi-natural habitats of Rumania (Banaszak & Manole, 1987) and Bulgaria (Banaszak & Dotchkova, unpublished data) (Table 2). Maximum daily densities of Apoidea (excluding Apis) in the period of optimum development of plants in Rumania and Bulgaria were as high as 3,000 individuals/ha. However, it should be stressed that also in colder climates the maximum daily density of wild Apoidea may be considerable in favorable conditions. For example, in a steppe reserve in Poland it exceeded 2,500 individuals/ha in April and May (Banaszak & Cierzniak, 1994a, b). In countries with milder climates, correspondingly higher densities are observed also on crops. For example, in alfalfa plantations bees were 17 times more numerous in Rumania than in Poland. Examples of densities of wild Apoidea in natural environments and on crops are given in Table 3.

Table 1. Percentage occurrence of honey bees, bumble bees and solitary bees crops in Poland (after Banaszak, 1983)

Crops	Honey-bee	Bumble-bee	Solitary bees
Winter rape	87.8	2.7	9.2
Sunflower	70.1	29.9	0.0
Flax	90.1	8.7	1.1
Alfalfa	98.0	1.5	0.5
Red clover	56.5	43.4	0.1
Yellow lupin	88.5	11.4	0.1
Buckwheat	95.3	4.5	0.2

Table 2. Average density (individuals/ha) of wild Apoidea in various biotopes of Rumania and Poland (after Banaszak & Manole, 1987)

Ecosystems	Rumania	Poland
Oak forest	682.5	245.8
Oak-hornbeam forest	1000.0	315.5
Steppe/Xerothermic sward	1159.5	805.0
Roadside	1420.0	318.5
Alfalfa plantation	3950.0	231.5
Sunflower plantation	150.0	5700.0

Table 3. Average density (individuals/ha) of wild Apoidea in grey dunes, grasslands and in alfalfa plantations

Country	Grey dunes	Grasslands	Alfalfa	Authors
Germany				
Spiekeroog (Jul)				
Juist (Jul)	550.0	650.0	-	Banaszak (unpubl. data)
Poland (Aug)	327.8	752	232	Banaszak (1983)
		1240	-	Banaszak & Cierzniak (1994)
Rumania		1285	3200-4700	Banaszak & Manole (1987)
Bulgaria		1500	1500-2500	Banaszak & Dotchkova (unpubl. data)
Greece				
Lefkada (Aug)	650.0	-	-	Banaszak (unpubl. data)

8.3 Pattern of Distribution of Apoidea in Agricultural and Natural Landscapes

Landscape ecology has become one of the crucial interdisciplinary fields of research in natural sciences, and an important basis for landscape management and conservation. Forman and Godron (1986) define landscape as an ecological system of a higher rank than the ecosystem; it is a fragment of the surface of the Earth composed of a number of ecosystem types which are regularly repeated in space and interact with one another. One of the major directions of landscape studies is research on landscapes which have been significantly changed by human activity, agricultural landscapes in particular.

The structure of agricultural landscapes is considered first of all as a system of patchy habitats, corridors and environmental barriers which form an ecological network within the matrix of arable fields (Forman & Godron, 1986). The surface area, shape, distribution, age, and level of disturbance of the individual patches and ecological corridors are of great importance for many groups of animals inhabiting the agricultural landscape.

Bees often penetrate several ecosystems and their migrations are conditioned by a number of environmental factors. Some bees nest in refuge ecosystems but forage for pollen and nectar in the neighboring habitats that provide more food. This phenomenon is favorable for farmers if they are able to locate seed crops appropriately. Among refuge habitats, the greatest biodiversity is observed in permanent natural or semi-natural communities such as forests and xerothermic grasslands (*e.g.* on sunny hillsides and roadsides). They provide shelter for various animals and support temporary habitats, which are equally important for insect survival, for example hedgerows, grassy field borders and roadsides. Bee diversity in such temporary habitats is as high as in natural biotopes (Banaszak, 1983). This is because ecological niches of grassy field borders and roadsides are similar to those of xerothermic grassland, while living conditions in hedgerows are similar to those in forests. The second type of landscape elements are habitats created by man: arable fields, meadows, orchards, etc. Due to the high degree of human pressure, they do not provide the majority of bees with suitable nesting sites. Only some bee species colonize these habitats, mainly plantations of perennial crops. Nevertheless, many crops are rich sources of pollen and nectar, so bees come there to forage for food. Thus, only a network of natural, semi-natural and agricultural ecosystems may support a maximally rich bee fauna in the agricultural landscape.

An analysis of bee diversity and density (Shannon-Weaver index, H') in Polish and Rumanian lowlands showed that there is an inverse interdependence between these parameters (Fig. 3). Habitats arranged according to an increasing diversity gradient - from cultivated areas to natural forest communities - are characterized by a decrease in bee density. Simplification of the structure of plant communities results in a lower bee diversity. For example, arable fields are characterized by an exceptionally high plant density but since they provide a homogeneous food source, a low number of bee species is associated with them. By contrast, the relatively high diversity of plants in natural ecosystems sustains a large number of animal species, despite lower plant densities. Furthermore, it was found that there is a positive correlation between the number of plant species and the number of bee species in the landscape (Fig. 4) (Banaszak, 1983, 1992a), and between the area covered in blooming plants and the density of Apoidea in refuge habitats of the agricultural landscape (Fig. 5) (Cierzniak, 1991, 1995). Similarly, a positive correlation between the extent of direct contact of a plantation with refuge areas and both the diversity (Fig. 6) and density (Fig. 7) of Apoidea in the plantation was detected (Banaszak & Cierzniak, 1994b).

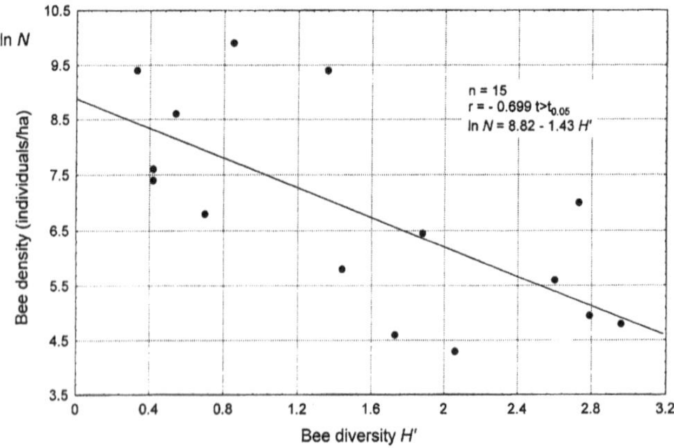

Figure 3. Relationship between the diversity (H') and density of Apoidea (ln N) in 15 habitats of an agricultural landscape (after Banaszak, 1983).

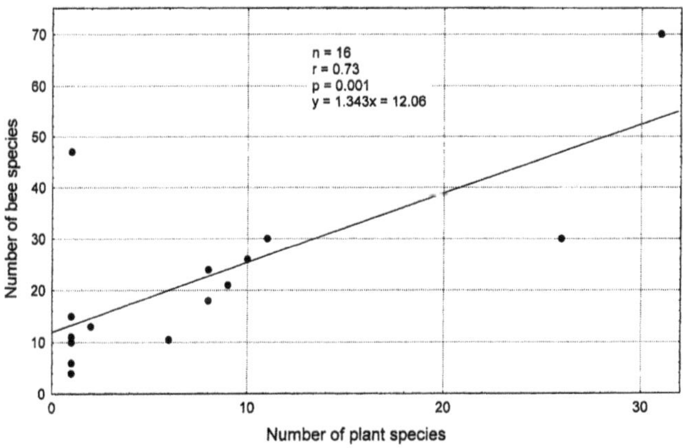

Figure 4. Relationship between the number of plant species and the number of bee species in the agricultural landscape (after Banaszak, 1983).

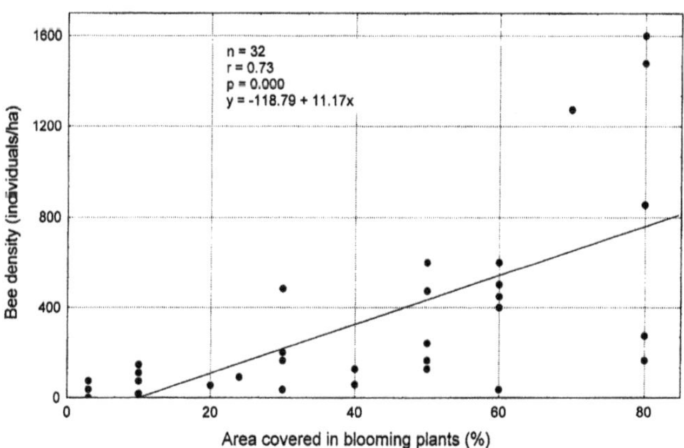

Figure 5. Relationship between the area covered in blooming plants and the density of Apoidea in refuge habitats of the agricultural landscape (after Cierzniak, 1991).

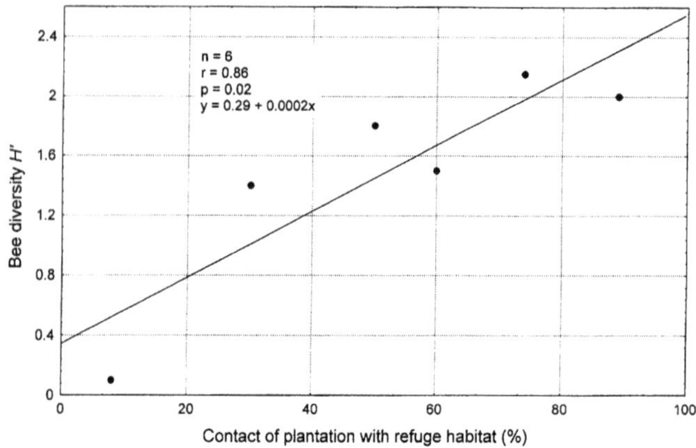

Figure 6. Relationship between the extent of direct contact of a plantation with a refuge habitat and the diversity of Apoidea (after Banaszak & Cierzniak, 1994b).

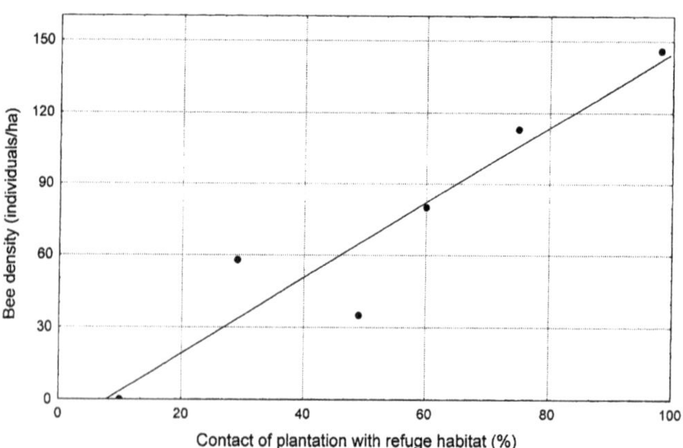

Figure 7. Relationship between the extent of direct contact of a plantation with a refuge habitat and the density of Apoidea (after Banaszak & Cierzniak, 1994b).

Studies of bee ecology in the agricultural landscape (Banaszak, 1983) suggest that there are some factors which compensate for the negative influence of land management on populations of wild bees. First of all, the agricultural landscape is a mosaic of arable land and refuges, the latter providing bees with suitable nesting sites. The second, additional factor favorable for bees, is introduction of large areas of bee-pollinated crops, such as oilseed rape, clover, alfalfa and sunflower, which substitute for wild forage plants. The proportion of arable land to refuges in the landscape, and the distribution of these landscape elements exert a decisive influence upon the survival of pollinators. Therefore, we should strive to protect the present resources of wild fauna by preserving the mosaic structure of the agricultural landscape.

8.4 Impact of the Structure of the Agricultural Landscape on Bee Diversity and Density

The rules governing the distribution of Apoidea which have been presented above have a practical significance for the shaping of the agricultural landscape and securing the stability of the landscape as a whole. Thus, we should aim at creating an appropriately diverse network of ecosystems: arable fields, meadows, hedgerows, roadsides, grassy field borders and barren ground. A solution to the difficult situation described at the beginning of this chapter may be an agricultural landscape reconstructed according to ecological principles, where pollinating insects would find everything they need.

A question arises: what should an appropriately constructed agricultural landscape look like to be able to secure the survival of an abundant and diverse bee fauna?

For three landscape types in western Poland, the total number of wild bees was calculated and compared with the share of refuge ecosystems in the landscape. On this basis, a landscape model represented by a curve in Fig. 8 was created (Banaszak, 1986). The rule governing the distribution of Apoidea reflected by this model suggests that the share of crop plantations in the agricultural landscape should not exceed 75% of the total. The rest should be refuges for the fauna. A proper ratio of these two types of biotopes and their distribution within the landscape are crucial for the survival of pollinators. Such a conclusion may be drawn from observations made in agricultural landscapes of central Poland (Banaszak, 1983) and Rumania (Banaszak & Manole, 1987), the two areas differing in climatic conditions and land use patterns.

Refuge areas are indispensable for the survival of bees and, therefore, are necessary for agriculture. Direct contact of a grassy roadside with an alfalfa plantation ensures the pollination of flowers of this plant. Patches of uncultivated ground surrounded by arable land are exposed to the influence of various crop husbandry practices and may be damaged by them. Thus, more stable and better protected "reserves" of the fauna are necessary to ensure bee survival. Refuge ecosystems exert an influence on all pollinating insects in the landscape. A comparison of two regions of Rumania - Bucharest (Calugareni-Comana) and the outskirts of Craiova - shows that the more varied landscape of Craiova suburbs supports a greater diversity and abundance of Apoidea than Calugareni, where the share of refuge habitats is low. Vast areas of monocultures create particularly stressful conditions for the fauna. Enlargement of arable fields to the maximum leads to the disappearance of bee refuge habitats. While recognizing the desire to increase agricultural productivity, one should remember that productivity of insect-pollinated plants depends also on the quality and level of pollination (Table 4).

Investigations conducted by Cierzniak (1995) in the area of Poland also suggest that the overall structure of the landscape shapes the natural resources of Apoidea. In agricultural landscapes of the lowlands of west Poland, numbers of wild bees were 3.6 times higher in a highly varied landscape (18% of refuge habitats), than in a landscape with a very low share of refuge habitats (2.8% of refuge habitats) (see Table 5).

Also Pawlikowski (1989) found that the diversity and density of Apoidea is higher when arable fields are divided into small patches. This is associated with the greater area of field borders and roads with roadsides, which provide sites for the development of this group of insects.

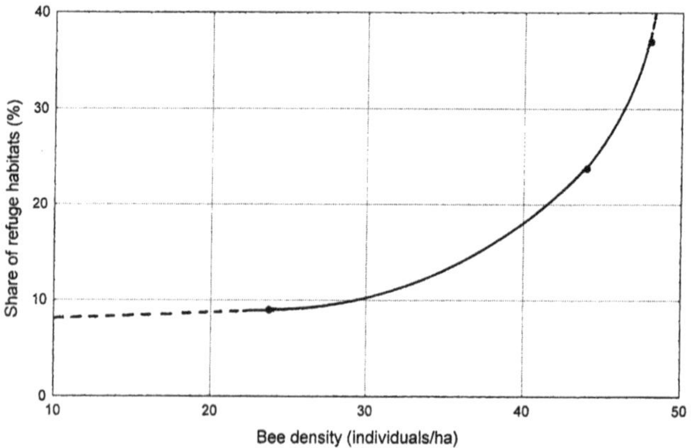

Figure 8. Dependence of bee density on the share of refuge habitats in agricultural landscapes.

8.5 Fauna Refuges in the Agricultural Landscape

Forests in many countries of Europe have been reduced to small fragments surrounded by intensively cultivated arable fields. In this way, human activity resulted in separation of habitats of many animals into small, isolated islands. These islands, together with hedgerows and grasslands, are refuges for the fauna and flora. The more intensive crop husbandry practices, the more important are the refuges for the survival of wild animals (Czarnecki, 1956; Banaszak, 1983; Opdam *et al.*, 1985; Dąbrowska-Prot, 1991, Loster, 1991; Loster & Dzwonko, 1988; Kozakiewicz & Szacki, 1987).

Table 4. Impact of agricultural landscape structure on the diversity and density of Apoidea (after Banaszak & Manole, 1987)

Parameter	Bucharest region	Craiova region
Type of landscape	Smaller share of refuge habitats.	Greater share of refuge habitats, particularly natural grasslands and forests.
Typical lowland landscape.	Varied land relief.	
Number of species	49	72
Percentage of the total number of bee species recorded in Rumania lowlands	46	68
Density of wild bees on alfalfa plantations (individuals/ha)	3200	4700

Table 5. Participation of families of Apoidea in communities of refuge habitats in complex and simplified agricultural landscapes (after Cierzniak, 1995)

	Complex landscape			Simplified landscape		
Bee family	Mean density (ind./ha)	%	Number of species	Mean density (ind./ha)	%	Number of species
Apidae	216.9	71.6	11	58.2	44.3	9
Halictidae	43.1	14.2	16	45.3	33.9	15
Andrenidae	19.4	6.4	10	22.3	17.0	7
Other*	23.4	7.7	18	5.7	4.9	6
Total	302.8	100.0	55	131.5	100.0	37

* - Collectidae, Melittidae, Megachilidae, Anthophoridae

In contrast to the short-term supplies of bee forage in crop monocultures, wild plants of the refuges provide bees with continuous supplies of pollen and nectar throughout the growing season thanks to a succession of flowering herbs, shrubs and trees (see Table 6).

Table 6. Succession of bee forage plants in a shelter belt (after Cierzniak 1995)

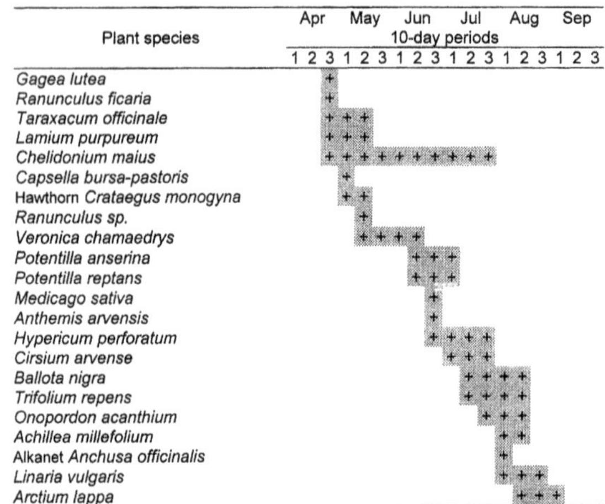

Plant species	Apr			May			Jun			Jul			Aug			Sep		
	10-day periods																	
	1	2	3	1	2	3	1	2	3	1	2	3	1	2	3	1	2	3
Gagea lutea			+															
Ranunculus ficaria			+															
Taraxacum officinale			+	+	+													
Lamium purpureum			+	+	+													
Chelidonium maius			+	+	+	+	+	+	+	+	+	+						
Capsella bursa-pastoris				+														
Hawthorn *Crataegus monogyna*				+	+													
Ranunculus sp.					+													
Veronica chamaedrys					+	+	+	+										
Potentilla anserina								+	+	+								
Potentilla reptans								+	+	+								
Medicago sativa									+									
Anthemis arvensis									+									
Hypericum perforatum									+	+	+	+						
Cirsium arvense										+	+	+						
Ballota nigra											+	+	+	+				
Trifolium repens											+	+	+	+				
Onopordon acanthium												+	+	+				
Achillea millefolium													+	+				
Alkanet *Anchusa officinalis*													+					
Linaria vulgaris													+	+	+			
Arctium lappa														+	+	+		

In patches of forest many species of plants grow providing bees with an abundance of forage, which is reflected in their high pollen and nectar production (see Table 7).

Table 7. Pollen and nectar productivity of some herbs commonly found in shelter belts (after Demianowicz, 1979; Demianowicz *et al.*, 1960; Jabłoński, 1968; Jabłoński *et al.* 1992; Jabłoński, 1994)

Plant species	Pollen and nectar productivity per 1 ha
Alkanet Anchusa officinalis	up to 461 kg nectar
Bindweed Convolvulus arvensis	10.2-55.1 kg pollen
Bird's-foot-trefoil Lotus corniculatus	26 kg nectar
Chicory Cichorium intybus	up to 100 kg nectar
Cornflower Centaurea cyanus	up to 95 kg pollen and 100 kg nectar
Dandelion Taraxacum officinalis	71-370.9 kg pollen and 25 kg nectar
Dead-nettle, Red Lamium purpureum	26-50 kg nectar
Dead-nettle, White Lamium album	15.2 kg pollen and 368-700 kg nectar
Horehound, Black Ballota nigra	40 kg nectar
Motherwort Leonurus cardiaca	300-400 kg nectar
Vetch, Crown Coronilla varia	5.6 kg pollen
Vetch, Fodder Vicia villosa	50 kg nectar
Viper's-bugloss Echium vulgare	182 kg nectar

Wild herbs are most numerous along the edges of forest patches. They are often accompanied by shrubs which are also rich sources of bee forage (see Table 8).

Table 8. Honey productivity of some tree species commonly found in shelter belts (after Demianowicz, 1979; Demianowicz *et al.*, 1960; Jabłoński, 1968; Jabłoński *et al.* 1992; Jabłoński, 1994)

Plant species	Honey productivity per 1 ha
Blackberry Rubus spp.	20 kg
False-acacia Robinia pseudoacacia	50-over 500 kg
Gooseberry Ribes grossularia	20 kg
Hawthorn Crataegus spp.	100-200 kg
Lime, Large-leaved Tilia platyphyllos	630 kg
Lime, Small-leaved Tilia cordata	1000 kg
Maple, Field Acer campestre	over 500 kg
Maple, Norway Acer platanoides	230 kg
Prunus divaricata	40 kg
Raspberry Rubus idaeus	100 kg
Willows and sallows Salix spp.	20-30 kg

Research carried out in the agricultural landscape of mid-west Poland (Cierzniak, 1995) shows that the extent of spatial isolation of refuge habitats is important for the survival of populations of individual bee species. In the studied types of agricultural landscape it was found that exchange between bee populations of refuges that are up to about 450 m apart is not limited by the high level of human interference in the arable land separating them. This also augments the availability of bee forage by allowing these insects to use the resources which are unevenly distributed in the spatially isolated refuge habitats. It was found that the diversity and density of wild bees is not dependent on the area of the refuge, but on the total area covered by communities of bee forage plants. In woodlots the crucial element increasing the diversity of bees is herb vegetation along their edges. Because of this, the magnitude of bee forage resources of these habitats depends mainly on their circumference, not on their surface area. Thus, to protect and enrich the bee fauna, it is best to promote narrow wooded shelter belts covering a relatively small area and having a long borderline where herb communities may develop (Fig. 9).

The development of herb communities along the edges of forest patches is conditioned by the closeness of crop plantations and on the applied husbandry practices (*e.g.* ploughing). If the border of the arable field lies very close to tree trunks, herb vegetation does not develop there because of shade and lack of space. Conversely, if the border lies outside the area shaded by trees, there is enough space for the development of herb communities. If there is an additional buffer zone between the forest patch and the field, for example a road or a ditch, herb and shrub communities are particularly rich. Such a situation was observed at the mouth of the Vistula River (Cierzniak, 1996). Shelter belts created in this area in 1964-1969 are usually located along roads or ditches, or are accompanied by strips of grassland. The space between the shelter belt and the field allows for the development of herb vegetation which attracts large numbers of bees. Along the edges of these shelter belts bee density was about 1400 individuals/ha. Where there was no such buffer zone, herb vegetation was scarce and bee numbers were low. Thus, the location of shelter belts along roads and drainage ditches, which was dictated by practical reasons, proved to be advantageous for bees, too.

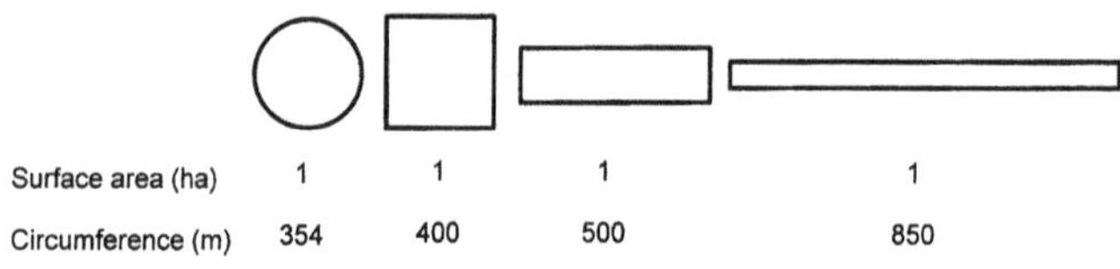

Surface area (ha)	1	1	1	1
Circumference (m)	354	400	500	850

Figure 9. Relationship between the surface area and the circumference of a woodlot depending on its shape (after Cierzniak, 1995).

The abundance of bee forage plants along the edges of larger forest patches attracts large numbers of bees (see Fig. 10). Inside large wood lots bees are usually less abundant than along their edges, mainly because of the shade. Nevertheless, when the trees are in flower, particularly in spring, bees can collect the pollen and nectar produced by bee-pollinated tree species. The quantity of bee forage produced by woody vegetation is then in proportion to the share of bee-pollinated trees and the area covered by the woodlot.

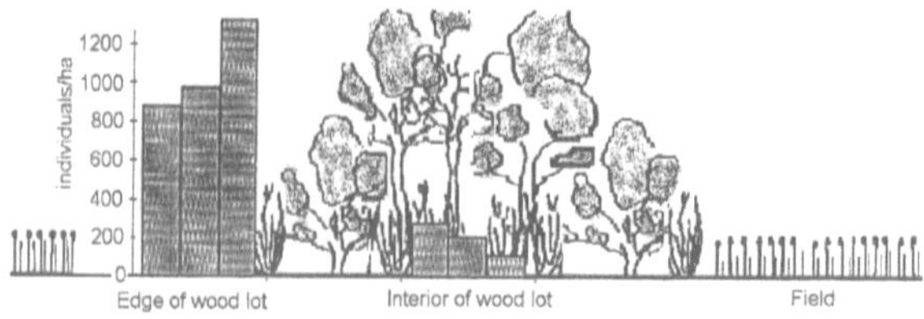

Figure 10. Mean density of Apoidea along the edges and inside a woodlot surrounded by arable fields (after Banaszak & Cierzniak, 1998).

Forest patches in the agricultural landscape are not only a rich source of food for bees but also provide wild species of Apoidea with suitable nesting sites. Forest patches and their edges are colonized by many species of bumble bees, Halictidae, Melittidae, as well as Colletes, Andrena, Hylaeus, Megachile, Osmia, Clisodon and Ceratina species.

In the above-mentioned investigations into the bee fauna of wood lots surrounded by arable land in the Puszcza Zielonka Landscape Park, it was found that six wood lots covering a total area of 1 ha, were inhabited by 96 bee species, which constitutes 25% of the fauna of bees in Poland (Banaszak *et al.*, 1996). Bee density may also be high in such habitats. For comparison, Table 9 presents data on the diversity and density of Apoidea in selected types of ecosystems.

Table 9. Density of Apoidea in shelter belts and other woodlots in agricultural landscapes as well as other natural and semi-natural ecosystems in various locations in Poland

Ecosystem type	Bee density (ind./ha)		No. of species	Author
Range	Mean	Range		
Shelter belts, n=3 (Gen. D. Chłapowski Landscape Park)	122.0-613.1	587.6	12-26	Banaszak, 1983; Cierzniak, 1995
Other wood lots, n=2 (Gen. D. Chłapowski Landscape Park)	986.0-1022.3	1004.1	14-29	Cierzniak, 1995
Other wood lots, n=6 (Puszcza Zielonka Landscape Park)	1147.6-347.9	746.0	20-60	Banaszak & Cierzniak, (1998)
Forests, n=3 (Wielkopolska National Park)	69.1-74.5	71.8	39-48	Banaszak, 1983; Banaszak & Cierzniak, 1995
Forests, n=4 (Wigry National Park)	42.0-406.0	180.5	59-135	Banaszak & Krzysztofiak, 1996
Xerothermic grassland, n=3	537.0-1314.4	980.0	59-135	Banaszak, 1983; Banaszak & Cierzniak, 1995; Banaszak & Krzysztofiak, 1996

8.6 Final Remarks and Conclusions

Bees (Apoidea) play a crucial role in farming as the insects that carry the main weight of plant pollination. Modern agriculture, particularly the use of herbicides and pesticides, other crop husbandry practices and establishment of extensive monocultures, contributed to the decline of the populations of these beneficial insects. At the same time, the need for the pollinating activity of bees is now higher than ever before. Paradoxically, the desire to increase seed production has resulted in a greater demand for pollinators. A solution to this difficult situation can be found in a common effort of scientists, bee-keepers and farmers. Attempts at breeding wild bees are promising. *Megachile rotundata* has been domesticated and used in alfalfa seed plantations. Attempts at breeding some bumble bees have also been successful. Research into domestication of other bee species, *e.g.* ground-nesting species, is underway but presents some major difficulties. At the same time, the work of many ecologists aims at improving the structure of the agricultural landscape to provide enough space for bees and other useful animals. Although in many cases nature can defend itself against human interference, it is our duty to help.

The above considerations show that possibilities of maintaining or even increasing the diversity and numbers of pollinating insects are linked with the landscape itself. Firstly, it is necessary to shape the agricultural landscape in a proper way, *i.e.* to make sure that the area covered by arable land does not exceed 75% of the total and the distribution of various types of ecosystems is appropriate. Secondly, the share of natural and semi-natural ecosystems should be large, as it increases the diversity of bees. Thirdly, it is important to protect fauna refuges, particularly national parks, nature reserves, landscape parks, etc. Not only should legal protection of these areas be provided, but also true protection, based on ecological rules of the functioning of the whole landscape. Finally, it is necessary to optimize the use of herbicides and pesticides. They should be applied by specialized groups of conscientious and professionally trained workers with high-quality equipment.

These are the necessary conditions which allow us to be quite optimistic about the future of farming and about bee resources. All in all, it must be emphasized that bee numbers need to be augmented and pollination should be treated as equivalent to the standard crop husbandry practices. Indeed, better pollination of arable crops means increasing food supplies for the human population.

8.7 References

Banaszak, J., 1980. Studies on methods of censusing the numbers of bees (Hymenoptera, Apoidea). *Pol. Ecol. Stud.* **6(2)**: 355-366.

Banaszak, J., 1983. Ecology of bees (Apoidea) of agricultural landscape. *Pol. Ecol. Stud.* **9(4)**: 421-505.

Banaszak, J., 1986. Impact of agricultural landscape structure on diversity and density of pollination insects. In: *Impacts de la structure des paysages agricoles sur la protection des cultures*, Poznań, 9-14 septembre 1985. Ed. INRA, Les Colloques de l, INRA, Paris, **36**: 75-84.

Banaszak, J., 1987. Fauna pszczół (Hymenoptera, Apoidea) Niziny Wielkopolsko-Kujawskiej na przestrzeni pół wieku. [Bee fauna of Wielkopolska-Kujawy Lowland during fifty years.] *Badania Fizjograf. Pol. Zach.*, **36, C**: 67-77.

Banaszak, J., 1989. Investigations on natural resources of pollinators. In: *Unconventional methods in lucerne breeding*. Proceedings Medicago sativa Working Group Meeting, Radzików, Sept. 12-17, 1988, pp. 27-29.2

Banaszak, J., 1991. A checklist of the bee-species (Apoidea) of Poland with remarks of their taxonomy and zoogeography. Acta Univ. Lodz., *Folia Zool. Anthr.* **7**: 15-66.

Banaszak, J., 1992a. Strategy for conservation of wild bees in an agricultural landscape. *Agric. Ecosystems Environ.* **40**: 179-192.

Banaszak, J. (Ed.), 1992b *Natural resources of wild bees in Poland*, Pedagogical Univ., Bydgoszcz, 174 pp.

Banaszak, J. (Ed.), 1995. Changes in fauna of wild bees in Europe. Pedagogical University, Bydgoszcz, 220 pp.

Banaszak, J., 1997. Local changes in the population of wild bees. I. Changes in the fauna ten years later. *Ochrona Przyrody*, **54**: 119-130.

Banaszak, J. & Cierzniak, T., 1994. Spatial and temporal differentiation of bees (Apoidea) in the forests of Wielkopolski National Park, Western Poland. *Acta Univ. Lodz., Fol. zool.* **2**: 3-28.

Banaszak, J. & Cierzniak, T., 1994a. Estimate density and diversity of Apoidea in steppe reserve "Zbocza Płutowskie" on the lower Vistula river. *Pol. Pismo Ent.* **63(3-4)**: 319-336.

Banaszak, J. & Cierzniak, T., 1994b. The effect of neighbouring environments and the acreage of the winter rapaseed plantations on the diversity and density of Apoidea (Hymenoptera). *Zeszyty Nauk. WSP, Stud. Przyr.* **10**: 25-38.

Banaszak, J. & Cierzniak, T., 1995. Ekonomiczne efekty zapylania roślin uprawnych przez pszczołę miodną i dziko żyjące pszczołowate (Apoidea). [Economical effects of arable crops pollination by honey bee and wild bees (Apoidea)]. *Kosmos* **44(1)**: 47-61.

Banaszak, J. & Cierzniak, T., 1998. Bees - Apoidea. (In:) Banaszak J. (Ed.), *Ecology of Forest Islands*. Pedagogical Univ., Bydgoszcz.

Banaszak, J, Cierzniak, T., Kaczmarek, S., Manole, T., Piłacińska, B., Ratyńska, H, Szwed, W. & Wiśniewski, H., 1996. Biodiversity of Forest Islands in an Agricultural Landscape. *Biul. Polish Acad. Sci., Biol. Sci.* **44,1-2**: 11-119.

Banaszak, J. & Krzysztofiak, A., 1996. The Natural Wild Bee Resources (Apoidea, Hymenoptera) of the Wigry National Park. *Pol. Pismo Ent.* **65**: 33-51.

Banaszak, J. & Manole, T., 1987. Diversity and density of pollinating insects (Apoidea) in the agricultural landscape of Rumania. *Pol. Pismo Ent.* **57(4)**: 747-766.

Ceballos, G., 1956. Catalogo de los himenopteros de Espana. Cousejo superior de Investigaciones cientificas, Madrid, p. 554.

Cierzniak, T., 1991. Wstępna ocena zgrupowań pszczół (Hymenoptera, Apoidea) w dwóch typach krajobrazu rolniczego [Preliminary estimate of bee communities (Hymenoptera, Apoidea) in two types of agricul tural landscape]. *Wiad. Entomol.* **10,4**: 70-76.

Cierzniak, T., 1995. The effect of a differentiation of an agricultural landscape on Apoidea communities. *Zesz. Nauk. WSP, Stud. Przyr.* **11**: 13-50.

Cierzniak, T., 1996. Wstępna ocena zadrzewień Żuław Gdańskich jako biotopu owadów zapylających [Preliminary evaluation of Żuławy Gdańskie afforestations as pollinating insect biotopes]. *Zesz. Nauk. WSP w Bydgoszczy, Studia Przyr.* **12**:75-86.

Czarnecki, Z., 1956. Materials for birds ecology nesting in the field afforestations. *Ekol. Pol. A*, **4(13)**: 379-417.

Dąbrowska-Prot, E., 1991: The role of forest island in the shaping of the structure and functioning of entomofauna in an agricultural landscape. *Ekol. pol.* **39(4)**: 481-516.

Demianowicz, Z,. 1979. Nektarowanie i wydajność miodowa Taraxacum officinale Web. [Nectar secretion and honey efficiency of Taraxacum officinale Web.]. *Pszczeln. Zesz. Nauk.* **23**: 97-103.

Demianowicz, Z., Hłyń, M., Jabłoński, B., Maksymiuk, J., Podgórska, J., Ruszkowska, B., Szklanowska, K. & Zimna, J., 1960. Wydajność miodowa ważniejszych roślin miododajnych w warunkach Polski (część I) [Nectar secretion and honey efficiency of important honey plants growing under Poland's conditions (Part I)]. *Pszczeln. Zesz. Nauk.* **10**: 87-94.

Forman, R.T.T. & Godron, M. 1986: *Landscape ecology*. New York, John Wiley and Sons.

Goldsmith, E & Hildyard, N. (Eds.), 1988. *The Earth Report. Monitoring the Battle for Our Environment.* Mitchell Beazley, London, 240 pp.

Jabłoński, B., 1968. Wydajność miodowa ważniejszych roślin miododajnych w warunkach Polski. Cz. IV [Nectar secretion and honey efficiency of important honey plants growing under Poland's conditions (Part IV)]. *Pszczeln. Zesz. Nauk.* **12, 3**: 117-125.

Jabłoński, B., 1994. Ogródek pszczelarski. Wyd. ISK Skierniewice, Puławy, 54 pp.

Jabłoński, B., Kołtowski, Z. & Dąbska, B., 1992. Nektarowanie i wydajność miodowa ważniejszych roślin miododajnych w warunkach polski - cz. VII [Nectar secretion and honey efficiency of important honey plants growing under Poland's conditions (Part VII)]. *Pszczeln. Zesz. Nauk.* **36**: 54-64.

Janzon, L.-Å., Svensson, B.G. & Erlandsson, S. 1991. Catalogus Insectorum Sueciae. Hypmenoptera, Apoidea. *Ent. Tidskr.* **112**: 93-99

Kocourek, M., 1989. Enumeratio insectorum Bohemoslovakiae [Check list of Czechoslovak Insects III (Hymenoptera)]. *Acta Faun. Entomol. Mus. Nat. Pragae* **19**: 173184.

Kozakiewicz, M. & Szacki. J., 1987. Drobne ssaki środowisk izolowanych, wyspy na lądzie czy tylko populacje wyspowe? *Wiad. Ekol.* **33**; 31-45.

Loster, S., 1991. Różnorodność florystyczna w krajobrazie rolniczym i znaczenie dla niej naturalnych i półnaturalnych zbiorowisk wyspowych. *Fragm. Flor. Geobot.* **36(2)**: 427-457.

Loster, S. & Dzwonko, Z. 1988. Species richness of small woodlands on the western Carpathian foothills. *Vegetatio* **76**: 15-27.

Monsevicius, V., 1995. A checklist of wild bee species of Lithuania with data to their distribution and bionomics. In: *New and rare for Lithuania insect species.* Institute of Ecology, Vilnius, pp. 7-145.

Opdam, P., Rijsdijk, G. & Hustings, F., 1985: Bird communities in small woods in an agricultural landscape: effects of area and isolation. *Biol. Conserv.* **34**: 333-352.

Osytśhnjuk, A.Z., Panfilov, D.V. & Ponomareva, A.A., 1978. Opredelitel nasekomych evropejskoj casti SSSR. Nadsem. Apoidea Pcelinyje, Izd. Nauka, Leningrad, **3**: 279-519.

Pawlikowski, T., 1989: Struktura zgrupowań dzikich pszczołowatych (Hymenoptera, Apoidea) z obszarów o różnych typach parcelacji powierzchni uprawnej [The structure of wild bee (Hymenoptera, Apoidea) communities from farming areas of different field sizes]. *Acta Univ. Nicolai Copernici* **33,70**: 31-46.

Pagliano, G., 1995. Checklist delle specie della fauna Italiana. Fasc. 106. Hymenoptera, Apoidea. Ministero dell 'Ambiente a Comitato Scientifico per la Fauna d'Italia, Ed. Calderini, Bologna. pp. 1-25.

Rasmont, P., Ebmer, A., Banaszak, J. & Van der Zanden, G., 1995. Hymenoptera Apoidea Gallica. Liste taxonomique des abeilles de France, de Belgique, de Suisse et du Grand Duche de Luxemburg. *Bulletin de la Société Entomologique de France* **100**: 1-98.

Tanacs, L., 1982. Untersuchung der blumenbesuchenden bienenformigen Insectenpopulation (Hymenoptera: Apoidea) auf dem Rasen-Okosestem der Bugacer Sandheide. *Folia Entomologica Hungarica*, **43(1)**: 179-190.

Vikberg, V., 1986. A checklist of aculeate Hymenoptera of Finland (Hymenoptera, Apocrita, Aculeata).*Notulae Entomologicae* **66**: 65-85.

Westrich, P., 1984. Kritisches verzeichnis der Bienen der Bundesrepublik Deutschland (Hymenoptera, Apoidea). *Courior Forschungeinstitute Senckenberg* **66**: 1-86.

CHAPTER 9

DIVERSITY AND MOVEMENT PATTERNS OF LEAF BEETLES (COLEOPTERA: CHRYSOMELIDAE) AND LEAFHOPPERS (HOMOPTERA: CICADELLIDAE) IN A HETEROGENEOUS TROPICAL LANDSCAPE

Implications for Redressing the Integrated Pest Management Paradigm

MICHAEL E. IRWIN
Department of Natural Resources and Environmental Sciences, University of Illinois at Urbana — Champaign, Urbana, IL

LOWELL R. NAULT
Department of Entomology, OARDC, The Ohio State University, Wooster, OH

CAROLINA GODOY
Instituto Nacional de Biodiversidad, Santo Domingo de Heredia, Heredia, Costa Rica

GAIL E. KAMPMEIER
Center for Economic Entomology, Illinois Natural History Survey, Urbana, IL

9.1 Introduction

The forests of Central America (Williams, 1994), as elsewhere in the tropics (Westman *et al.*, 1989; Bennett, 1990), are being cut down at an alarming rate (Meyer & Turner II, 1994). Much of the land resulting from this activity is cleared and converted to unimproved pasture, forming a patchy landscape mosaic of forest and glade (Franklin & Forman, 1987) in a number of life zones throughout the region.

On warm, summer mornings and afternoons, especially when the sun is at the proper angle and the air nearly still, multitudes of insects can be seen flitting between these tropical forests and pastures. Although the insects appear to concentrate their flight along and through forest gaps, the aggregate direction of movement is not obvious. Biotic interchanges between habitats have been under investigation by a number of workers during the past decade, mostly addressing movement of insects on farmlands and between cultivated cropland and hedgerows (Wiens *et al.*, 1993, 1997; With *et al.*, 1997; numerous chapters in this book).

According to Burel and Baudry (1995b), landscape heterogeneity plays an important but poorly understood role in the dispersal of organisms between habitats. Differences in forest patch size and isolation account for nearly all of the variability in dispersal success,

B. Ekbom, M. Irwin and Y. Robert (eds.), Interchanges of Insects, 141-168

with larger patches having significantly greater exchanges of dispersing organisms (Gustafson & Gardner, 1996). Using simulation models, Stamps *et al.* (1987) demonstrated a positive correlation between edge permeability and the tendency of a disperser to cross the boundary and emigrate once it reached the edge of a habitat patch.

Some workers have focused on aerially mobile insects between agricultural and hedgerows or wildlands (*e.g.*, Saville *et al.*, 1997, Jeanneret & Charmillot, 1995), but seldom has the magnitude, dynamics, and impact of these biotic interchanges been documented in the literature. Because many of these observations include interchanges of pests, vectors, pollinators, beneficial organisms and other insects of potential economic importance, movement patterns need to be better understood (Stinner *et al.*, 1983; Fry, 1995) so that resulting impacts can be included and mitigated when devising and deploying crop, forest, and landscape management strategies.

Economic impact to any system can be high when interchanges of specific classes of pests are involved. From the standpoint of potential impact, perhaps the most important are those that transmit plant viruses (Irwin & Nault, 1996). Viruses are often severely debilitating to crop or pasture systems being invaded and, once introduced, are extremely difficult to control. Vectors are consequential because they carry viruses into a system, initiating epidemics, and once within the system, spread the pathogens from plant to plant. The mobility of vectors is thus a driving force behind virus epidemics, and the intensity and severity of epidemics is proportional to vector movement activity (Irwin & Ruesink, 1986).

This study provides a definitive step towards documenting the diversity and movement patterns of insects across system boundaries (*i.e.*, ecotones) in tropical landscapes. We explore temporal and directional components of this interchange by examining the diversity and movement of two families of insects - leaf beetles (Insecta: Coleoptera: Chrysomelidae) and leafhoppers (Insecta: Homoptera: Cicadellidae) - within and between tropical forests and largely unimproved pastures in Costa Rica. Both families contain numerous species that transmit viruses to wild and cultivated plants (Nault, 1997).

9.2 Study Sites

To develop a more comprehensive understanding of local movement patterns of these vector groups, three study sites were selected in Costa Rican landscapes with forest and adjacent pasture in distinct and contrasting life zones (Fig. 1). The three sites differ greatly in rainfall intensity (Fig. 2) and in the quantity of solar radiation (Fig. 3) but not ambient temperature (Fig. 4). Each site has two additional characteristics in common, a well demarcated ecotone between forest and pasture and a professional or paraprofessional entomologist stationed at the site who was responsible for the collection and initial curation associated with the project. The three sites chosen for this study form a rough transect across Costa Rica (Fig. 1):

* Nuñez in the tropical dry forest life zone,
* Montezuma in the tropical moist forest, premontane belt life zone, and
* Cocorí in the tropical wet forest life zone.

Nuñez [85° 08' W, 10° 20' S]. At about 70 m above sea level on the western lower slope of Cerro Eskameca, Estación Experimental Enrique Jímenez Nuñez is situated in the lowlands of south central Guanacaste Province (Fig. 1). This once dry tropical forest consists primarily of remnant stands of dry deciduous forest, interspersed with improved and unimproved pasture and experimental crop land. Rainfall is seasonally heavy and interrupted by a long, dry period during which many of the forest trees lose their leaves. Nuñez was the most disturbed of the three sites studied. At the Nuñez study site, monitoring of vectors occurred for 41 days, between 13 June and 23 July 1993. Collecting occurred during the short rainy season.

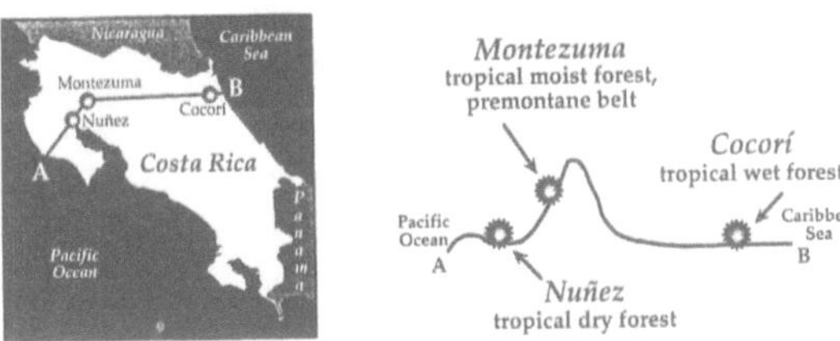

Figure 1. Map of Costa Rica showing the positions of the study sites through which a line transect (A—B) extending from the Pacific Ocean to the Caribbean Sea has been drawn (left). A relief diagram along the line transect (A—B) provides the relative topography and positioning of the three study sites (right).

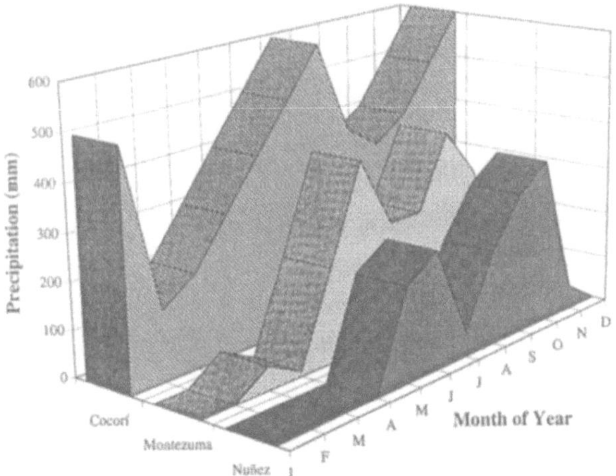

Figure 2. Average monthly precipitation (mm) at the three study sites (after Barrantes *et al.*, 1985).

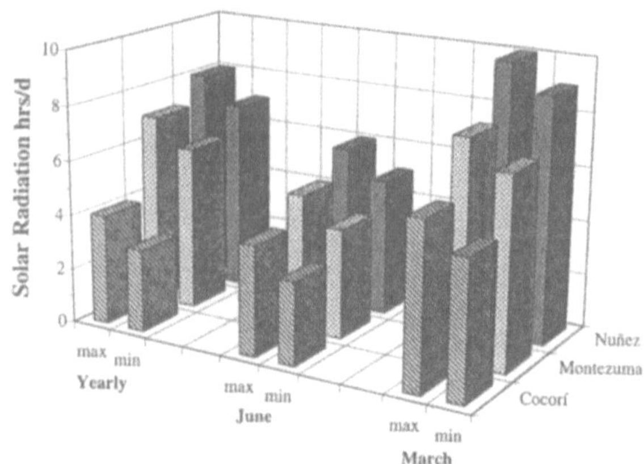

Figure 3. Solar radiation (hours per day) at the three study sites averaged over the year and during the extreme months of June and March (after Barrantes *et al.* 1985).

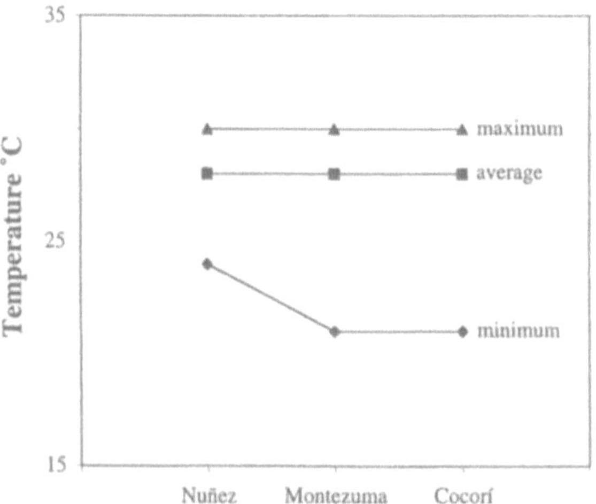

Figure 4. Maximum, minimum, and average temperatures at the three study sites (after Barrantes *et al.* 1985).

Montezuma [85° 06' S, 10° 42' W]. Granadero Montezuma is situated in the premontane belt at about 480 m above sea level on the northwest facing slope of Cerro Montezuma, Cordillera de Tilarán. Even though seasonally heavy, rain falls almost every month; thus, the dry period is not intense. However, an extended period of strong east to west winds is characteristic of the site during the end of the long rainy period continuing into the period with the least amount of precipitation. The Granadero contains a dairy ranch and scattered pastures, fruit tree orchards, coffee plantations, and crop land. It is bordered to the south, east, and west by a large, contiguous tropical moist forest. The ecotone between forest and pasture was little disturbed by human activity. At the Montezuma study site, monitoring of vectors occurred for 33 days, between 27 October and 28 November 1993. Collecting occurred at the end of the long rainy season when winds were strong.

Cocorí [85° 45' S, 10° 20' W]. Cocorí is situated between the Tortuguera and the Suerte Rivers at <30 m above sea level in the tropical wet forest of eastern Costa Rica. The site is located at the southern limit of a nearly pristine forest that stretches northward to and beyond the Nicaraguan border. The terrain at the site is rather cut up, with recently cleared but unimproved pastures scattered among remnant tracts of forest. The ground is saturated much of the year. Cattle grazing is common, with small, irregular patches of ground in crops such as taro that are common to the humid lowland tropics. The collecting area at Cocorí was along the edge of a pristine forest and was the least disturbed of the three sites. At the Cocorí study site, vectors were monitored for 44 days, between 04 November and 17 December 1993. Collecting occurred during the height of the long rainy season.

9.3 Insect Monitoring, Processing, Data Management, and Analyses

Six-meter, bidirectional Malaise traps (Fig. 5) (sold by J.W. Hock, Inc., c/o Debby H. Focks <jwhock@vector.net>, P.O. Box 12852, Gainesville, FL 32604) monitored these vector groups of insects at the three study sites. Each six-meter trap captured flying and walking insects approaching from the two long sides of the traps. Each side of a Malaise trap had an exposed opening 4.50 m long and 1.45 m high with a central baffle, giving it a trapping area of approximately 6.5 m^2. The traps were positioned so that the bottom of their central baffles made continuous contact with the ground. The traps were constructed of a neutral gray mesh fabric (5.8 x 6.3 lines per cm), which minimized both attraction and avoidance by the insects. At each of the three study sites, three Malaise traps were placed by us (MEI, LRN) parallel to each other and with the edge of the forest, one about 75 m into the forest, one about 75 m into the pasture, and one at the boundary between forest and pasture.

Some specimens of leaf beetle and leafhopper taxa[1] occasionally fly at an elevation greater than the catching height of the Malaise traps and thus fail to enter the traps, while a few of those that enter also exit the traps instead of finding their way into the collecting chambers. Thus, trap catches reflect somewhat biased abundances and are probably somewhat taxon biased as well; however, our daytime observations suggest that most specimens of our target taxa fly low enough to enter the traps and, once having entered,

[1] The terms "taxa" and "taxon" refer to the lowest levels of determination obtained for specimens; this was often at the species level, but the determinations could be made only to the genus level in a number of cases for specimens caught during the course of this study.

find their way to the collecting chambers. If these observations hold for night fliers as well, most specimens can be assumed to have entered the traps and were funneled into the collecting chambers, thus providing reasonable estimates of abundances and species richness during the time the traps were operated.

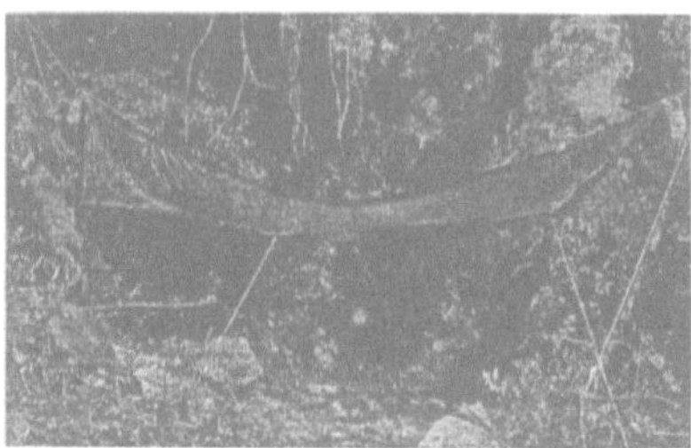

Figure 5. An ordinary six-meter bidirectional Malaise trap set up in the field.

The traps within the forest and pasture were "ordinary" bidirectional Malaise traps (Gressitt & Gressitt, 1962) that captured and funneled insects from both trapping directions into two cyanide-laced collecting chambers, one at the upper corner of each end of the trap. Samples from the two corner chambers on each of these traps were pooled. At each site, the forest trap was placed in a 5- to 6- m wide gap that provided a natural pathway between the denser inner forest and the open pasture. The pasture trap, held up by two poles set into the ground or hung between two widely spaced trees, was placed in an area that had a diversity of low growing plant species.

A "migration" Malaise trap (Walker, 1978) of exactly the same size and construction as the ordinary traps described above, but with a different method of separating the catch, was set at the boundary or ecotone of each site. At each of the upper corners of the migration traps were two collecting chambers, one that captured specimens entering from the forest side and the other for those entering from the pasture side. The trap catches from both corners that captured insects from the forest side were pooled and, similarly, the two corner chambers that captured specimens from the pasture side were pooled. Thus, each of these pooled samples contained the specimens that entered the Malaise from a single side.

At the Nuñez study site, the traps were gathered once a day, usually at sunrise (0600 h). At Montezuma and Cocorí, the insect samples were removed from the collecting chambers twice a day, at sunrise (0600 h) and sunset (1800 h), allowing additional data on flight diel periodicity (day vs. night) to be gathered. The samples were taken to a laboratory where the Chrysomelidae and auchenorrhynchus Homoptera were removed and mounted. The specimens were then sent to the Instituto Nacional de Biodiversidad (INBio) for further curation and identification. One of us (CG) was responsible for the curatorial process and the determinations at INBio. Most of the specimens captured during this study are housed in the INBio insect collection, Apartado Postal 22-3100, Santo Domingo de Heredia, Heredia, Costa Rica; e-mail: < cgodoy@rutela.inbio.ac.cr >. Two sets of vouchers exist, one with INBio, the other with the Illinois Natural History Survey (607 E. Peabody, Champaign, IL 61820, USA).

Data associated with each specimen were logged into the Biodiversity Information Management System developed and used by INBio. Those data (summarized in Table 1) were then compiled by INBio and sent to the University of Illinois at Urbana — Champaign where they were transferred to a FileMaker®Pro database for manipulation. Compiled data were analyzed for habitat and directional flight preferences using Microsoft®Excel for chi-square and G-tests. Taxon diversity measures, particularly the log series index α as a richness weighted measure of diversity and the Shannon Diversity index, the corrected H', as an evenness weighted measure of diversity, compared by t-tests, were calculated using formulae provided by Magurran (1988) and Krebs (1989).

9.4 Biodiversity Perspectives

Traps were monitored for 118 cumulative days at the three sites during the study. This resulted in the capture of 10,654 specimens of interest to us, 718 of which were leaf beetles and the remainder Auchenorrhyncha. These specimens represented 143 taxa (the specimens were identified to species where possible and to genus where species-level determinations were not possible). Within the Auchenorrhyncha, 8,694 specimens belonged to the family Cicadellidae. This study focuses on the abundance and temporal distribution of the taxa within the families Chrysomelidae and Cicadellidae, both rich in species that transmit plant viruses.

Biodiversity Among Sites. When comparing sites, samples from all three traps (six trapping sides) at each site were pooled. The family Chrysomelidae was consistently trapped during the study at all sites. A total of ten families within Auchenorrhyncha were also trapped during this study (Table 1), but taxa within the family Cicadellidae greatly outnumbered all other Auchenorrhyncha. Together, the nine traps collected an average per day of 80 specimens of both Chrysomelidae and Cicadellidae. Numbers caught differed among the sites, with the variation among sites much greater for Cicadellidae than Chrysomelidae. Nuñez had a mean catch of 4.9 Chrysomelidae and 75.2 Cicadellidae per day, while Montezuma had per day catches of 5.5 Chrysomelidae and 7.5 Cicadellidae. Cocorí, the site with the highest numbers on a daily basis, had mean daily catches of 7.7 leaf beetles and 121.9 leafhoppers. Overall, these specimens belonged to 74 taxa in 8 subfamilies of the Cicadellidae and another 41 taxa in 6 subfamilies of the Chrysomelidae (Table 1).

The trend of greater catch abundance in the wetter sites was consistent for the Chrysomelidae and held for the Cicadellidae for all sites except Montezuma, where the catch was lower on a daily basis than at the other two sites (Table 2). The low trap catches of leafhoppers at Montezuma were likely the result of trapping during the windy season. We suspect that leafhoppers flew minimally during that time because of strong and nearly continuous winds flowing east to west through the mountain pass and across the Montezuma site. Nonetheless, we are confident that most taxa within the Cicadellidae were trapped during the sampling interval at Montezuma because the cumulative taxon curve seems to have plateaued (Fig. 6). In fact, the cumulative number of taxa of Chrysomelidae and Cicadellidae approached plateaus through sampling effort (trapping days) at all sites (Fig. 6), suggesting that the sampling effort at each site was sufficient to have captured most of the taxa within the two families of insects that would normally enter the traps during the trapping season.

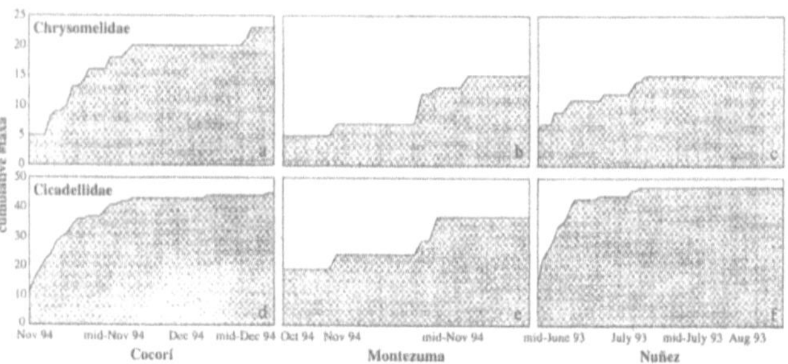

Figure 6. Leaf beetle (Coleoptera: Chrysomelidae) and leafhopper (Homoptera: Cicadellidae) taxa captured by Malaise traps at the three study sites, Cocorí, Montezuma, and Nuñez. Leaf beetle taxa (Figs 6a-6c) and leafhopper taxa (Figs. 6d-f) accumulated through sampling effort (number of days of trapping): 6a, Chrysomelidae at Cocorí; 6b, Chrysomelidae at Montezuma; 6c, Chrysomelidae at Nuñez; 6d, Cicadellidae at Cocorí; 6e, Cicadellidae at Montezuma; 6f, Cicadellidae at Nuñez.

Two biodiversity indices were calculated, α and the Shannon Diversity index or corrected H' (Magurran, 1988), the former biased towards richness (*i.e.*, number of taxa), the latter somewhat biased towards evenness (*i.e.*, less of a dominance of a few species in a sample). The α or log series calculations suggest that leaf beetles were richest at Cocorí, followed by Nuñez, and least rich at Montezuma. The exact reverse of this was shown for the leafhoppers, which were richest at Montezuma, followed by Nuñez, and least rich at Cocorí (Table 2).

Tests of significance (chi-square and G-test) for the corrected H' values among sites independently for the Chrysomelidae and Cicadellidae revealed that the only site comparison that had significantly different H' indices for Chrysomelidae was Nuñez vs. Montezuma ($P < 0.01$). For the Cicadellidae, however, the differences of corrected H' for Nuñez vs. Cocorí and Montezuma vs. Cocorí were highly significant ($P < 0.001$) (Table 2).

These two indices suggest that the diversity of Chrysomelidae and Cicadellidae may be negatively correlated; *i.e.*, where leaf beetles are more diverse, leafhoppers appear less so, and *vice versa*. Furthermore, within the environments we sampled, the higher the precipitation in the area, the greater the diversity of Chrysomelidae and the less diverse the Cicadellidae.

Table 1. Listing of the Chrysomelidae (Insecta: Coleoptera) and auchenorrhynchus Homoptera (Insecta) taxa captured during the study using Malaise traps, with a notation of the sites where the taxa were collected

Order	Family	Subfamily	Taxon	Cocorí	Montezuma	Nuñez
Coleoptera	Chrysomelidae	Alticinae	*Acanthonycha* sp.	x		x
			Alagoasa sp.			x
			Asphaera nobilitata (Fabricius)	x		
			Ayalaia minor Bechyne & Bechyne	x		
			Centralaphthona sp.	x		
			Chalatenanganya ?quadrifida			x
			Chalatenanganya quadrifida Bechyne & Bechyne		x	
			Coroicona sp.	x		
			Dinaltica sp.	x	x	
			Diphaulaca sp.		x	
			Glenidion sp.	x		x
			Heikertingerella sp.	x		x
			Heikertingeria sp.	x		
			Monoplatina sp.	x	x	
			Omophoita aequinoctialis (Linnaeus)	x	x	
			Parasyphrea sp.	x		
			Plectotetra sp.		x	
			Stegnea sp.	x		
			Systena sp.	x		
			Varicoxa sp.	x		
			Walterianella venustula (Schaufuss)		x	
		Chrysomelinae	*Platyphora petulans* (Stål)	x		

Table 1. Continued.

Order	Family	Subfamily	Taxon	Cocorí	Montezuma	Nuñez
Coleoptera		Eumolpinae	*Allocolaspis* sp.	x		
			Allocolaspis submetallica (Jacoby)		x	
			Antitypona sp.	x	x	x
			Brachypnoea sp.	x	x	x
			Caryonoda sp.	x		
			Cayetunya sp.			x
			Chalcophana mutabilis Harold		x	
			Cilaspis sp.			x
			Colaspis femoralis (Olivier)			x
			Habrophora sp.			x
			Metexyonycha tridentata Jacoby		x	
			Percolaspis sp.	x		
			Phanaeta sp.		x	
			Rhabdopterus sp.		x	
			Spintherophyta sp.	x	x	
			Typophorus sp.	x		x
		Galerucinae	*Masurius* sp.	x	x	
		Hispinae	*Chalepus horni* Baly		x	
		Megascelinae	Megascelis sp.			x
Homoptera (Auchenorrhyncha)	Cercopidae		*Aeneolamia albofasciata* (Lallemand)	x		x
			Aeneolamia postica Walker	x		
			Cephisus sp.		x	

Table 1. Continued.

Order	Family	Subfamily	Taxon	Cocorí	Montezuma	Nuñez
Homoptera			*Microsangarne* sp.	x		
(Auchenorrhyncha)			*Ocoaxo* sp.	x		
			Prosapia ca. *bicincta* (Say)		x	
			Prosapia plagiata (Distant)			x
			Prosapia simulans (Walker)	x		
			Zulia sp.	x		
			Zulia vilior (Foxler)	x		
	Cicadellidae	Agalliinae	*Apogonalia* sp.	x		
			Apogonalia fractinota (Fowler)			x
			Beirneola sp.	x		
			Carneocephala sp.		x	
			Dilobopterus sp.	x		
			Draeculacephala sp.	x	x	x
			Erythrogonia sp.	x	x	
			Graphocephala sp.	x		
			Graphocephala ca. *redacta* (Fowler)			x
			Graphocephala coccinea (Fowler)		x	
			Hortensia similis (Walker)	x	x	x
			Ladoffa sp.	x		
			Ladoffa sannionis Young		x	
			Macunolla ventralis (Signoret)	x		
		Coelidiinae	*Pilosana* sp.	x		
			Pilosana gratiosa (Spangberg)	x		

Table 1. Continued.

Order	Family	Subfamily	Taxon	Cocorí	Montezuma	Nuñez
Homoptera (Auchenorrhyncha)		Deltocephalinae	*Bulclutha* sp.	x	x	x
			Chlorotettix sp.	x		x
			Chlorotettix emarginatus Baker	x		x
			Chlorotettix minimus Baker	x	x	x
			Dalbulus sp.			x
			Graminella sp.		x	
			Ileopeltus tethys (Van Duzee)		x	x
			Iowanus sp.			x
			Ollarianus sexmaculatus Linnavuori			x
			Osbornellus sp.	x	x	x
			Osbornellus affinis (Osborn)	x	x	x
			Osbornellus blantoni Linnavuori			x
			Planicephalus flavicosta (Stål)	x	x	x
			Sanctanus fasciatus (Osborn)			x
			Scaphytopius sp.		x	x
			Spangbergiella vulnerata (Uhler)			x
			Tropicanus flectus DeLong	x	x	x
		Gyponinae	*Acusana* sp.	x		
			Curtara sp.	x		x
			Curtara objecta (Fowler)	x		x
			Gypona sp.		x	
			Gyponana sp.		x	
			Gyponana? sp.	x	x	x
			Hecalapona sp.	x	x	

Table 1. Continued.

Order	Family	Subfamily	Taxon	Cocorí	Montezuma	Nuñez
Homoptera			*Polana* sp.	x	x	x
(Auchenorrhyncha)			*Polana* ca. *unca* DeLong & Freytag	x	x	x
			Ponana sp.	x	x	
			Prairiana or *Gyponana* sp.		x	x
			Scaris sp.	x		
	Cicadellidae	Neocoelidiinae	*Chinaia* sp.	x		
		Typhlocybinae	*Alconeura* sp.	x	x	x
			Diceratalebra sp.		x	x
			Dikraneura sp.	x	x	x
			Empoasca sp.	x		
			Erabla sp.		x	x
			Erythroneura sp.			x
			Erythroneura? sp.			x
			Habralebra sp.		x	x
			Joruma sp.	x		x
			Omegalebra sp.	x	x	x
			Parallaxis sp.			x
			Parallaxis guzmani (Baker)			x
			Parallaxis ornata Osborn	x		x
			Protalebrella brasiliensis (Baker)		x	x
			Rabela sp.			x
			Rhabdotalebra sp.			x
			Trypanalebra maculata (Baker)			x

Table 1. Continued.

Order	Family	Subfamily	Taxon	Cocorí	Montezuma	Nuñez	
Homoptera	Cicadellidae	Xestocephalinae	*Portanus minor* Kramer	x	x		
(Auchenorrhyncha)			*Xestocephalus* sp.	x	x	x	
			Xestocephalus desertorum (Berg)		x	x	
			Xestocephalus luridus Linnavuori	x			
	Cicadidae		*Pacarina* sp.			x	
			Proarna sp.			x	
	Cixiidae		*Bothriocera* or *Bothrioceretta* sp.	x			
			Pintalia sp.	x	x		
			Rhamphixius sp.			x	
	Delphacidae		*Caenodelphax teapae?* (Fowler)		x		
	Derbidae		*Anotia* sp.	x	x	x	
			ca. *Anotia* sp.	x			
			Cedusa sp.	x	x	x	
			Patara sp.	x		x	
			Persis sp.			x	
			Sayiana sp.			x	
			Shellenius sp.			x	
			ca. *Shellenius* sp.			x	
	Dictyopharidae		*Lappida* sp.				x
	Fulgoridae		*Cladodiptera* or *Cladyphe* sp.		x		
	Issidae		*Picumna* sp.	x		x	
	Nogodinidae		*Biolleyana* sp.	x			

Table 2. Abundance (n), taxon richness (taxa), and biodiversity indices (α and Shannon) of Chrysomelidae and Cicadellidae, comparing the three study sites

Chrysomelidae	sites	n	taxa	α	corrected H'	variance H'
	Nuñez	200	15	4.4	2.025	0.0039
	Montezuma	181	15	3.9	1.792	0.0044
	Cocorí	337	24	5.9	1.946	0.0055
	t-test	df	t	*P*		
	Nuñez vs. Montezuma	376	2.6	**		
	Nuñez vs. Cocorí	533	0.8	n.s.		
	Montezuma vs. Cocorí	498	-1.5	n.s.		
Cicadellidae	**sites**	**n**	**taxa**	**α**	**corrected H'**	**variance H'**
	Nuñez	3091	40	6.5	2.476	0.0004
	Montezuma	249	34	10.6	2.498	0.0069
	Cocorí	5363	38	5.5	1.426	0.0005
	t-test	df	t	*P*		
	Nuñez vs. Montezuma	279	-0.3	n.s.		
	Nuñez vs. Cocorí	8211	34.8	***		
	Montezuma vs. Cocorí	278	12.5	***		

n.s. = not significant ($P > 0.05$); ** = $P \leq 0.01$; *** = $P \leq 0.001$

Biodiversity Within Sites. Many more specimens of both families were captured in the pastures than in the forests, regardless of site. As an extreme example, at Cocorí there were more than 13 times more specimens of Cicadellidae caught by the pasture trap than by the forest trap. Although this general trend held for the two habitats, the magnitude of the differences was usually considerably less (Table 3). The trend of greater abundance in the pasture was generally consistent with the number of taxa of both vector groups in the two habitats. The number of specimens captured per taxon by forest traps was less than that captured by pasture traps except for two taxa, both of which were leaf beetles at Montezuma. Taxa captured by pasture traps, however, were always less than twice those captured in the forest (Table 3).

Diversity indices α and corrected *H'* (Chapter 2, Magurran, 1988) were calculated separately for leaf beetles and leafhoppers captured by the forest and pasture traps at each site. The α index for the Cicadellidae is nearly double that of the Chrysomelidae in both habitats and across all sites. This index was also slightly higher for Chrysomelidae in the forest habitat than in the pasture, regardless of site. The highest α index in our study (α = 8.7) was calculated for leafhoppers in pasture catches at Montezuma. Contrary to the trend in the leaf beetles α diversity, this α index was higher for leafhoppers captured by the pasture traps at both Montezuma and Nuñez, although at Cocorí, the index was slightly higher for the taxa in the forest than the pasture (Table 3).

The values for the Shannon Diversity index for forest and pasture catches were compared at each site using a *t*-test (Magurran, 1988). This index always indicated less dominance by a few taxa in the forest than in the pasture, except for Cicadellidae at Nuñez, where the indices were equal. For Chrysomelidae, the difference between catches in the forest and pasture was significant at Cocorí ($P < 0.01$) and Montezuma ($P < 0.05$), but not at Nuñez ($P > 0.05$). For Cicadellidae, the difference was significant only at Montezuma ($P \approx 0.01$).

Although fewer individuals and taxa of Chrysomelidae and Cicadellidae were captured in the forest habitat, the Shannon Diversity index (corrected *H'*) was significantly higher in half of the cases. This suggests that catches in the forest were less dominated by a few species than those in the pasture. At the same time, the α index suggests that the forest habitat is richer in Chrysomelidae but that the pasture habitat is often richer in Cicadellidae.

Table 3. Abundance (n), taxon richness (taxa), and biodiversity indices (α and Shannon) of leaf beetles and leafhoppers, comparing forest and pasture habitats at each of the three study sites

Group	Site	Habitat	n	taxa	α	corrected H'	variance H'	df	t	*P*
Chrysomelidae	Nuñez							62	0.157	n.s.
		Forest	27	9	4.7	1.546	0.0285			
		Pasture	69	11	3.7	1.512	0.0178			
	Montezuma							41	1.852	*
		Forest	21	7	3.7	1.473	0.0198			
		Pasture	40	5	1.5	1.155	0.0096			
	Cocorí							80	2.704	**
		Forest	40	9	3.6	1.673	0.0149			
		Pasture	247	15	3.5	1.277	0.0065			
Cicadellidae	Nuñez							486	0.000	n.s.
		Forest	351	27	6.8	2.35	0.0000			
		Pasture	1857	39	7.1	2.35	0.0000			
	Montezuma							180	2.346	**
		Forest	55	16	7.6	2.257	0.0095			
		Pasture	129	24	8.7	1.882	0.0162			
	Cocorí							404	0.633	n.s.
		Forest	357	23	5.4	1.387	0.0076			
		Pasture	4874	31	4.5	1.33	0.0005			

9.5 Flight Activity Patterns

Because the Malaise trap samples were gathered at 0600 h and 1800 h at Cocorí and Montezuma, night catches (between 1800 h and 0600 h) could be compared with day catches (between 0600 h and 1800 h). Only activity patterns within sites were compared. Sixteen taxa were cumulatively abundant enough at one of the two sites to be compared. Where five or more specimens were caught during the night and daytime, the data were analyzed by both chi-square and G-tests. These analyses are invalid when actual numbers drop below five for night or day catches. When this occurred, the symbol "†" was inserted beside the appropriate figures in Table 4. In all instances, differences in magnitude of day and night catches for such taxa were so great that patterns were easily discernible.

The Montezuma site contained four taxa within the Chrysomelidae and two in the Cicadellidae that were caught in sufficient numbers to be compared. This is in contrast with Cocorí, which had sufficient data to compare diel activity for two taxa within the Chrysomelidae and eight within the Cicadellidae. The analyses provided strong evidence ($P < 0.001$) that some taxa were active during the day while others were active at night (Table 4). It must be remembered that crepuscular fliers are active during the time when the traps were changed and thus are distributed in both night and day samples. One would expect them to show no day/night preference and be indistinguishable from those taxa that were active during both the day and night. This set of taxa has been designated "none" (Table 4), as having no day or night flight activity preference.

At Montezuma, the four leaf beetle taxa included one that was decidedly more active at night, one that was caught more frequently during the day, and two that were not significantly more active during one time period than the others. Both leafhopper taxa were active during the night. At Cocorí, the two leaf beetle taxa were active during the day. Six of the eight leafhopper taxa were active at night, while the other two were not significantly more active in either of the two time periods. None of the Cicadellidae that were analyzed proved to be active only during the day. One of the five leaf beetle taxa was a night flier, three were day fliers, and two had no discernible periodicity preference. One taxon, *Brachypnoea* sp. (Chrysomelidae: Eumolpinae), was collected in sufficient numbers at both sites for analysis. At the Cocorí site, where 146 specimens were captured, flight activity was decidedly during the day. At the Montezuma site where 31 specimens were captured, no significantly discernible day/night flight pattern was uncovered even though more beetles were collected during the day than night (Table 4). These data suggest that although many leaf beetle taxa are day fliers, leafhopper flight activity seems predominantly a nighttime phenomenon. Very few studies have examined flight periodicity in the Cicadellidae. Nontheless, a few studies have been undertaken. For example, *Graminella nigrifrons* and three *Dalbulus* species were found to be crepuscular fliers. They leave the plant canopy at dusk and remain in flight during the warmer, early evening hours (Rodriguez *et al.*, 1992, Lopes *et al.*, 1995, Taylor *et al.*, 1993).

Table 4. Flight timing (day vs. night) of leaf beetle and leafhopper taxa at the Cocorí and Montezuma study sites

Family	Subfamily	Taxon	n	day	night prefrence	*P*
COCORÍ						
Chrysomelidae	Eumolpinae	*Percolaspis* sp.	30	26	4	day †
		Brachypnoea sp.	146	144	2	day †
Cicadellidae	Deltocephalinae	*Chlorotettix minimus*	236	38	198	night***
		Tropicanus flectus	187	7	180	night***
	Agallinae	*Agallia panamensis*	306	136	170	none n.s.
	Gyponinae	*Curtara* sp.	270	31	239	night***
	Cicadellidae	*Hortensia similis*	104	59	45	none n.s.
		Plesiommata corniculata	146	10	136	night ***
	Xestocephalinae	*Xestocephalus* sp.	3214	350	2864	night ***
		Xestocephalus luridus	426	57	369	night ***
MONTEZUMA						
Chrysomelidae	Eumolpinae	*Allocolaspis submetallica*	32	1	31	night †
		Antitypona sp.	38	31	7	day ***
		Brachypnoea sp.	31	19	12	none n.s.
	Alticinae	*Dinaltica* sp.	47	25	22	none n.s.
Cicadellidae	Deltocephalinae	*Balclutha* sp.	37	1	36	night †
	Gyponinae	*Polana* sp.	71	2	69	night †

n.s. = not significant ($P > 0.05$); *** = $P \leq 0.001$ according to both chi-square and G-tests
† = one of the numbers was too low to test by either chi-square or G-test.

9.6 Abundances Within and Movement Between Habitats

Abundances within habitats. Whether a taxon was more abundant in the forest or pasture was determined by comparing the cumulative catches from the ordinary six-meter Malaise traps placed within the two habitats at each site. Comparisons based on cumulative catches were made using both chi-square and G-tests. Because these analyses were invalid when actual numbers were fewer than five for any given category (*e.g.* habitat preference for forest vs. pasture), the symbol "†" was inserted next to the taxa where applicable (Table 5). In cases where the catch in one was very low and the other very high, the higher catch was assumed to indicate the preferred habitat. However, where the catches were not obviously lopsided in the two habitats, it was impossible to refute the null hypothesis that the catches were equal. This happened infrequently for the catches in the two habitats. In cases where no decision could be made, the data are presented in Table 5 but not discussed further.

Of the taxa having more than 5 individuals in any trap over the trapping season, the richest site, with 16 taxa, was Nuñez, followed by Cocorí with 14, and Montezuma with 6. Of these, 7 taxa were leaf beetles and 27 were leafhoppers, both spread among the three sites. For the Chrysomelidae, forest or pasture catches were too small to be tested in five of eight instances, although, for *Antitypona* sp. at Nuñez, ten times more specimens were captured in the forest (n = 10) than in the pasture (n = 1). Specimens of the same genus were captured at Montezuma in significantly ($P < 0.05$) greater numbers in the pasture (n = 15) than in the forest (n = 6) (Table 5). The only other leaf beetle taxon that had a decided preference for habitat was *Brachypnoea* sp. at Cocorí, where a highly significant ($P < 0.001$) preference for pastures (n = 135) over forest (n = 9) existed. That same taxon in Montezuma was not tested but three times more specimens were caught in the pasture (n = 13) than in the forest (n = 4).

Leafhoppers were caught in significantly higher numbers in the pasture than forest in most cases (Table 5). At Montezuma, these differences were less striking than in the other two sites, although 57 specimens of *Polana* sp. were caught in the pasture and only 1 in the forest. At Nuñez, catches were usually significantly higher in the pasture, but in one instance catches were higher in the forest. Only two taxa were not tested: catches of *Chlorotettix emarginatus* were higher in the pasture (n = 128) than in the forest (n = 4); catches of *Parallaxis ornata* were much higher (n = 75) in the forest than in the pasture (n = 1). In three cases, no preference could be detected. At Cocorí, leafhoppers were overwhelmingly more abundant in the pasture. None were captured in higher numbers in the forest, but two taxa were caught in such low numbers that they were not tested.

On the one hand, some leaf beetles were captured more frequently in the forest while others were found more abundantly in the pasture samples. On the other hand, almost all leafhopper taxa were caught more frequently in the pasture. This does not suggest that they were more abundant in one habitat than the other. It does indicate, however, that the density of leafhopper specimens actively flying was higher in the pasture than in the forest. Because the flora in the forests is both diverse and widely scattered, its herbivorous denizens such as leafhoppers are presumably also more widely dispersed. This may partly explain the higher numbers of leafhoppers caught in the pasture.

Movement between habitats. Directional flight — whether a taxon was flying from the forest to the pasture or from the pasture to the forest — was detected by comparing the catches of the two sides of the migration Malaise trap placed at the ecotone of and parallel to the boundary where forest and pasture meet at each site. Comparisons based on cumulative catches were made using both chi-square and G-tests. Because these analyses were invalid when actual numbers were fewer than five for any given category

Table 5. Forest and pasture habitat catches and directional flight preferences of leaf beetle and leafhopper taxa at the three study sites

MONTEZUMA

Family	Subfamily	Taxa	n	forest	pasture	preference *P*	>>f	>>p	Direction
Chrysomelidae	Eumolpinae	*Allocolaspis submetallica*	22	4	1	none n.s.	17	0	from forest †
		Antitypona sp.	38	6	15	pasture *	15	2	from forest †
		Brachypnoea sp.	31	4	13	pasture †	11	3	from forest †
	Alticinae	*Dinaltica* sp.	47	2	0	none n.s.	41	4	from forest †
Cicadellidae	Deltocephalinae	*Balclutha* sp.	37	6	12	pasture *	18	1	from forest †
	Gyponinae	*Polana* sp.	74	1	57	pasture †	14	2	from forest †

NUÑEZ

Family	Subfamily	Taxa	n	forest	pasture	preference *P*	>>f	>>p	Direction
Chrysomelidae	Eumolpinae	*Antitypona* sp.	37	10	1	pasture †	14	12	none
Cicadellidae	Deltocephalinae	*Chlorotettix emarginatus*	211	4	128	pasture †	74	5	from forest ***
		Doleranus tethys	569	20	488	pasture ***	59	2	from forest †
		Osbornellus sp.	26	16	6	forest *	4	0	†
		Osbornellus affinis	158	9	76	pasture ***	58	15	from pasture ***
		Osbornellus blantoni	133	34	39	none n.s.	33	27	none
	Gyponinae	*Curtara objecta*	479	12	283	pasture ***	158	26	from forest ***
	Cicadellidae	*Sibovia occatoria*	35	6	14	none n.s.	15	0	from forest †
		Graphocephala sp. nr. *redacta*	31	8	17	none n.s.	6	0	†
	Typhlocybinae	*Omegalebra* sp.	222	33	179	pasture ***	5	5	none
		Parallaxis ornata	99	75	1	forest †	16	7	none
		Rabela sp.	116	17	97	pasture ***	2	0	†
		Erabla sp.	54	14	33	pasture **	6	1	†

Table 5. Continued.

NUÑEZ

Family	Subfamily	Taxa	n	forest	pasture	preference *P*	>>f	>>p	Direction
Cicadellidae		*Erythroneura* sp.	34	6	21	pasture **	6	1	†
	Xestocephalinae	*Xestocephalus* sp.	166	12	85	pasture ***	69	0	from forest †
		Xestocephalus desertorum	364	9	143	pasture ***	204	8	from forest ***

COCORÍ

Family	Subfamily	Taxa	n	forest	pasture	preference *P*	>>f	>>p	Direction
Chrysomelidae	Eumolpinae	*Brachypnoea* sp.	146	9	135	pasture ***	2	0	†
		Percolaspis sp	30	12	14	none n.s.	2	2	†
		Spintherophyta sp.	20	5	4	none n.s.	6	5	none
Cicadellidae	Deltocephalinae	*Chlorotettix minimus*	236	5	228	pasture ***	3	0	†
		Tropicanus flectus	187	13	173	pasture ***	1	0	†
	Agallinae	*Agallia panamensis*	306	28	274	pasture ***	3	1	†
	Gyponinae	*Curtara* sp.	272	12	259	pasture ***	0	1	†
		Hecalapona sp.	36	6	27	pasture ***	2	1	†
	Cicadellidae	*Ladoffa* sp.	14	5	5	none †	4	0	†
		Tylozygus geometricus	31	2	8	none †	16	5	from forest *
		Hortensia similis	104	3	100	pasture ***	1	0	†
		Plesiommata corniculata	146	13	133	pasture ***	0	0	†
	Xestocephalinae	*Xestocephalus* sp.	3214	242	2942	pasture ***	22	8	from forest **
		Xestocephalus luridus	426	0	426	pasture ***	0	0	†

In both chi-square and G-tests, * = $P < 0.05$, ** = $P \leq 0.01$, *** = $P \leq 0.001$; † = not enough specimens present in one or more traps to test by chi-square or G-test; >>f = from forest; >>p = from pasture.

(*e.g.* directional movement from forest vs. from pasture), the symbol "†" was inserted next to the taxa where applicable (Table 5). In cases where the catch in one was very low but in the other was very high, the higher catch was assumed to indicate the preferred movement direction. However, where the catches were not obviously lopsided in the two sides of the migration trap, it was impossible to refute the null hypothesis that the catches were equal, a situation that occurred frequently for the catches on the forest and pasture sides of the migration trap. In cases where no decision could be made, the data are presented in Table 5 but not discussed further.

The Chrysomelidae were collected in such low numbers in the migration traps at all sites that only in one instance was the difference in directional catch tested, *Antitypona* sp. at Nuñez, where no directional preference was detected (Table 5). Even though no tests could be performed, catches of the four leaf beetle taxa captured in Montezuma appeared to be considerably higher on the forest side of the traps, suggesting that there was a movement preference from forest to pasture. The specimens collected at Cocorí were so few that no inferences could be made, except in the case of *Spintherophya* sp., which showed no directional preference.

The Cicadellidae were also collected in low numbers at Montezuma and Cocorí in at least one side of the migration traps. For the taxa from Montezuma, a strong tendency was found for specimens to be captured on the forest side of the migration trap (Table 5). This trend continued for the leafhopper taxa at Cocorí, with only two taxa in sufficient numbers on both sides of the trap to be tested. In those instances, there was a significant preference for the forest side of the traps. Only at Nuñez were there sufficient species in sufficient numbers to detect a trend. Here too, overwhelmingly higher catches were found in the forest sides of the traps. Only in one instance, *Osbornellus affinis*, did we detect a strong tendency towards higher catches in the pasture sides of the traps. Only three taxa (*Osbornellus blantoni*, *Omegalebra* sp., and *Parallaxis ornata*) showed no preference towards one trap side or the other.

Movement and aggregation. Habitat catches and directional preferences of nine of the most commonly collected taxa are graphically and proportionally displayed (Fig. 7). Three leaf beetle taxa, one each from the three study sites, and six leafhopper taxa, one from Montezuma, one from Cocorí, and four from Nuñez, are presented to provide a panorama of taxon abundances and movement patterns encountered during this study. Because the migration traps accumulated specimens on a per-trapping-side basis, specimens captured in the ordinary traps in the forest and pasture were halved, and the four trapping sides were summed. A percentage of the catch from each side was calculated by dividing that side's catch into the summed catch. Therefore, the plotted numbers represent percentages of specimens of a taxon collected per side of a six-meter Malaise trap. Note that the abundances in the pastures were often greater than those in the forests and that the forest sides of the migration traps often captured higher numbers of individuals. These tendencies were pervasive throughout the study.

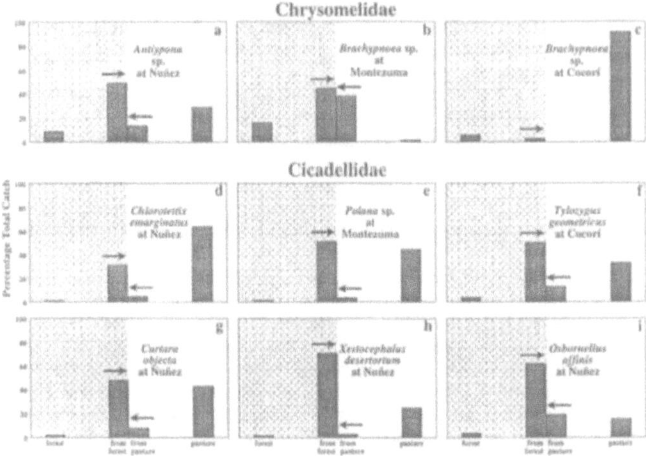

Figure 7. Three leaf beetle (Coleoptera: Chrysomelidae) taxa and six leafhopper (Homoptera: Cicadellidae) taxa captured by Malaise traps at the three study sites, Cocorí, Montezuma, and Nuñez. The migration traps collected specimens on a per-trapping side-basis, while the ordinary traps captured specimens from both trapping sides. To normalize the catches, the collections from the ordinary traps were halved to provide an average catch per trapping side. Numbers presented are percentages of the specimens captured in the combined four trapping sides (one in the forest, one at the forest side of the ecotone, one at the pasture side of the ecotone, and one in the pasture) over the entire trapping seasons. Figure 7, a-c: Leaf beetle taxa; 7a, *Antitypona* sp. at Nuñez; 7b, *Brachypnoea* sp. at Montezuma; 7c, *Brachypnoea* sp. at Cocorí. Figure 7, d-i: Leafhopper taxa, 7d, *Chlorotettix emarginatus* at Nuñez; 7e, *Polana* sp. at Montezuma; 7f, *Tylozygus geometricus* at Cocorí; 7g, *Curtara objecta* at Nuñez; 7h, *Xestocephalus desertortum* at Nuñez; 7i, *Osbornellus affinis* sp. at Nuñez.

The evidence thus accumulated during this study suggests that:

- leaf beetles and leafhoppers are often richer (greater number of taxa) and almost always more evenly distributed (ratios of numbers of individuals per taxon vary less) in forests than pastures,
- leaf beetles are often more active during the daytime while leafhoppers are almost always more active at night,
- most leaf beetle and leafhopper taxa are considerably more abundant in pastures, even though a few appear to be more abundant in forests, and some are more or less equally distributed in forests and pastures, and
- most taxa of leafhoppers and leaf beetles preferentially move from the forest to the pasture.

This leads to the formulation of the hypotheses that:

- most specimens of leaf beetle and leafhopper taxa are distributed sparsely in the forest habitat and move preferentially towards the pasture habitat, some during the night and others during the day,
- once having arrived in the pasture habitat, the specimens of these taxa remain there and do not preferentially move back to the forest, suggesting that
- the pasture habitat attracts and then acts as a sink that retains specimens of both leaf beetle and, most especially, leafhopper taxa.

9.7 Implications for Redressing the Integrated Pest Management (IPM) Paradigm

In practice, most agriculturally oriented IPM programs integrate appropriate control tactics into a strategy to manage a target pest in a given field. Their aim is often to maintain the pest below established economic or action thresholds and their methods are usually reactive or therapeutic (*i.e.*, action to cure an acute or chronic pest problem, see Pedigo, 1996). Several IPM programs are broader based and include both therapeutic and preemptive or preventative (*i.e.*, action against a pest before injury occurs, see Pedigo, 1996) tactics to manage multiple pests under monoculture or polyculture regimes. Because the management of one or a few pests usually influences the dynamics of other biotic aspects of the regime, field-level integration across multiple pests is being incorporated into a number of management strategies. Occasionally, the concept of IPM encompasses multiple pests, multiple tactics, multiple crops, and multiple fields. This higher level integration only infrequently has been undertaken, but the theoretical framework is in place (Kogan, 1998).

The pest management paradigm has different scales of resolution. However, as currently practiced, IPM is still rather narrowly applied to resolving pest problems in portions of agricultural systems, specific niches of urban systems (*e.g.* structural or garden), target pests in forestry systems (*e.g.* spruce budworm), or specific pests (*e.g.* gypsy moth) in natural systems. The ability to predict or forecast biotic events, such as movement (Hoy *et al.*, 1990), and perturbations that can impact a given system is an important preventative element in any IPM arsenal. It is this knowledge that allows the development and implementation of tactics needed to mitigate the ability of a pest or group of pests to disrupt the system. The current focus thus fails to cover the diversity and complexity encompassing any one of the systems mentioned, let alone combinations of systems.

Very seldom has the concept of IPM extended beyond the boundaries of the domain being managed, even when that domain is broadly defined to include agriculture, forestry, urban systems, or human health problems. Nonetheless, at least from the agricultural perspective, the idea of managing more than the area under cultivation is not new; the notion of manipulating field margins and hedgerows to maintain, shelter, and enhance the production of natural enemies has long been recognized (Wratten & van Emden, 1995) and has received considerable attention (*e.g.* Harwood *et al.*, 1994). Field margin management that enhances biological control is sometimes not utilized because it may also intensify pest buildup, for example, in situations where crop pests carry out key parts of their life cycles such as aggregation and mating of the European corn borer in grassy field margins (Showers *et al.*, 1976, DeRozari *et al.*, 1977). Thus, in a limited fashion, pest control strategies have begun to take into account management of non-cultivated lands adjacent to crops.

Area-wide pest management (Kogan, 1994, 1995), an important concept that had its written origins in the late 1970s or early 1980s (Knipling, 1980), theoretically focuses on key pests and encompasses their geographical home ranges (Schneider, 1989), assuring that the origins of these pests, be they local or migratory, form a part of the management domain. The concept is so new that it is only now beginning to be put into practice (Kogan, 1994, 1998). Thus, the theoretical framework of IPM embraces the management of key taxa that are active in both cultivated and wild settings. However, no IPM framework seems explicitly to address these interactive aspects across cultivated and natural landscape mosaics for multiple pests.

Inter-field movement of insects is difficult to quantify (Turchin *et al.*, 1991). Nonetheless, Duelli, *et al.* (1990) have evidence suggesting that almost all arthropod taxa move between agricultural and semi-wild ecosystems. Movement has considerable consequences to both systems (Burel & Baudry, 1995a). Our study provides supportive evidence that there is a continual and pervasive interchange of biota between managed (pasture) and unmanaged (tropical forest) lands. The studied biota represent groups of insects that contain important vectors of plant pathogens capable of severely decreasing crop harvests. In our study, leafhopper movement was almost entirely unidirectional from the forest to the pasture. This was also often true for leaf beetle movement. That the cultivated area, in this case a pasture, acted as a sink for most species of leafhoppers and some species of leaf beetles suggests that once the vectors enter a glade such as a pasture or perhaps also a cropped field, they move freely within it, providing an excellent mechanism for explosive plant viral epidemics. Without taking the wild lands and their biota into account, a functionally integrated pest management program would be nearly impossible to develop; management of viral epidemics transmitted by leafhoppers in such a tropical setting would be close to hopeless.

Kogan (1998), in his historical treatment of IPM, divided its practice into three levels. Level I involves the deployment of control strategies for single species at the field scale. Level II integrates multiple pests with multiple control tactics at the community scale. Level III integrates multiple pests and multiple control tactics at the cropping systems and agroecosystem scales.

This study indicates that there may be a need for an even higher order level in Kogan's series, one which integrates multiple pests and multiple control tactics at the landscape scale, a scale that embraces agroecosystems and relevant natural or wild systems. Given that mitigating the ability of a pest to negatively impact a system and that pest species move among systems, forecasting the movement and dynamics of these pests ought to be an important aspect of the management paradigm. Although it can be argued that Level III encompasses this aspect, only approaches that are of an area-wide nature currently incorporate wild areas into the management plan, and these plans focus on single key pests. By augmenting the current IPM paradigm so that cultivated and natural landscape mosaics are an integral part of the management strategy, we believe that IPM will retain and amplify its predictivity, allowing the preventative mitigation of potential pests. Thus, we argue here that the IPM paradigm needs to be redressed to assure the incorporation of wild or natural ecosystems into an overall management strategy.

Acknowledgments

Rodrigo Gámez, Director of INBio, was in charge of Costa Rican activities associated with this project. Jorge A. Jímenez located the site at Cocorí and arranged for INBio to process specimens from Cocorí and Montezuma. Alvaro Jenkens, owner, and Marco Chaves, manager, assisted by allowing one of our students to live at and place traps on the grounds of Granadero Montezuma during the collecting interval there. Raphael Mena, Director of the Enrique Jímenez Nuñez Agricultural Experiment Station, assisted the project by allowing access to the collecting site on the station and sanctioning the use of critical facilities. Frank D. Parker, then Leader of the U.S. Department of Agriculture's Screwworm Project in Costa Rica, kindly provided lodging and laboratory facilities under his control at the Nuñez station for initial sorting of samples from that location.

Richard Allan, Cornell University and former student at the University of Illinois at Urbana — Champaign, was responsible for monitoring the traps at Nuñez and Montezuma. Elias Rojas, an INBio parataxonomist, was responsible for this task at Cocorí. Will Flowers, Florida A & M University, Tallahassee, FL, kindly identified the Chrysomelidae. The Cicadellidae and other Auchenorrhyncha were mostly identified by one of us (CG), with help for the subfamilies Typhlocybinae and Gyponinae by Paul H. Freytag, University of Kentucky, Lexington, KY, for the superfamily Fulgoroidea by Lois O'Brien, Florida A & M University, Tallahassee, FL, and for the family Cercopidae by Vinton Thompson, Roosevelt University, Chicago, IL. Specimens collected at Nuñez were sent to Illinois, where they were processed by Erin Leslie, Leslie Marsh, and Honghong Zhang. All other material was processed at INBio. Mark A. Metz kindly calculated biodiversity indices, *t*-tests, G-tests, and chi-square analyses, and Audrey Fisher provided provisional analyses. We are extremely grateful to these people and their institutions for their multifaceted contributions toward the realization of this effort.

The project was supported in large part under Grant No. 10.236, "Natural systems as reservoirs of vectors of agriculturally important plant viruses and mycoplasms," Program in Science and Technology Cooperation, Office of the Science Advisor, U.S. Agency for International Development. USAID, INBio, University of Illinois at Urbana — Champaign, Illinois Natural History Survey, and The Ohio State University are gratefully acknowledged for financial and logistical support throughout the project.

9.8 References

Barrantes, J.A. (Coordinador), Liao, A. & Rosales., A., 1985. *Atlas Climatológico de Costa Rica.* Ministerio de Agricultura y Ganadería, Instituto Meteorológico Nacional. Proyecto MAG-CORENA. San José, Costa Rica.

Bennett, A.F., 1990. Habitat corridors and the conservation of small mammals in a fragmented forest environment. *Landscape Ecology* **4**: 109-122.

Burel, F. & Baudry, J., 1995a. Farming landscapes and insects. In: Glen, D.M., Greaves, M.P. & Anderson, H.M. (Eds.), *Ecology and Integrated Farming Systems*. John Wiley & Sons Ltd., NY, 203-220 pp.

Burel, F. & Baudry, J., 1995b. Species biodiversity in changing agricultural landscapes: A case study in the Pays d'Auge, France. *Agriculture, Ecosystems and Environment* **55**: 193-200.

DeRozari, M.B., Showers, W.B. & Shaw, R.H., 1977. Environment and the sexual activity of the European corn borer. *Environ. Entomol.* **6**: 658-665.

Duelli, P., Studer, M., Marchand, I. & Jakob, S., 1990. Population movements of arthropods between natural and cultivated areas. *Biological Conservation* **54**: 193-207.

Franklin, J.F. & Forman, R.T.T., 1987. Creating landscape patterns by forest cutting: Ecological consequences and principles. *Landscape Ecology* **1**: 5-18.

Fry, G., 1995. Landscape ecology of insect movement in arable ecosystems. In: Glen, D.M., Greaves, M.P. & Anderson, H.M. (Eds.), *Ecology and Integrated Farming Systems*. John Wiley & Sons Ltd., NY, pp. 177-202.

Gressitt, J.L. & Gressitt, M.K., 1962. An improved Malaise trap. *Pacific Insects* **4**: 87-90.

Gustafson, E.J. & Gardner, R.H., 1996. The effect of landscape heterogeneity on the probability of patch colonization. *Ecology* **77**: 94-107.

Harwood, R.W.J., Hickman, J.M., MacLeod, A., Sherratt T.N. & Wratten, S.D., 1994. Managing field margins for hoverflies. In Boatman, N. (Ed.), *Field margins: Integrating agriculture and conservation*. British Crop Protection Council monograph 58. Farnham, UK, pp. 147-152.

Hoy, C.W., McCulloch, C.E., Sawyer, A.J., Shelton, A.M. & Shoemaker, C.A., 1990. Effect of intraplant insect movement on economic thresholds. *Environ. Entomol.* **19**: 1578-1596.

Hunt, R.E. & Nault, L.R., 1991. Roles of interplant movement, acoustic communication, and phototaxis in mate-location behavior of the leafhopper *Graminella nigrifrons*. *Behav. Ecol. Sociobiol.* **28**: 315-320.

Irwin, M.E. & Nault, L.R., 1996. Virus/vector control. In: Persley, G.J. (Ed.), *Biotechnology and Integrated Pest Management*. Biotechnology in Agriculture No. 15, CAB International, London, pp 304-322.

Irwin, M.E. & Ruesink, W.G., 1986. Vector intensity: a product of propensity and activity. In: McLean, G.D., Garrett, R.G. & Ruesink W.G., (Eds.), *Plant Virus Epidemics: Monitoring, Modelling and Predicting Outbreaks*. Academic Press, Sydney, pp. 13-33.

Jeanneret, P. & Charmillot, P.J., 1995. Movements of tortricid moths (Lep. Tortricidae) between apple orchards and adjacent ecosystems. *Agriculture, Ecosystems and Environment* **55**: 37-49.

Knipling, E.F., 1980. Areawide pest suppression and other innovative concepts to cope with our more important insect pest problems. In: The Minutes of the 54th Anniversary Meeting of the National Plant Board. Sacramento, CA, pp. 68-97.

Kogan, M., (Ed.), 1994. *Areawide management of the codling moth: Implementation of a comprehensive IPM program for pome fruit crops in the western U.S.* Integrated Plant Protection Center, Corvallis, OR, 158 pp.

Kogan, M., 1995. Areawide management of major pests: Is the concept applicable to the *Bemisia* complex? In: Gerling, D. & Mayer, R.T. (Eds.), *Bemisia: Taxonomy, Biology, Damage, Control and Management*. Intercept, Andover, UK, pp. 643-657.

Kogan, M., 1998. Integrated pest management: Historical perspectives and contemporary developments. *Ann. Rev. Entomol.* **43**: 243-270.

Krebs, C.J., 1989. *Ecological Methodology*. Harper and Row, NY, 654 pp.

Lopes, J.R.S., Nault, L.R. & Phelan, P.L., 1995. Periodicity of diel activity of *Graminella nigrifrons* (Homoptera: Cicadellidae) and implications for leafhopper dispersal. *Ann. Entomol. Soc. Am.* **88**: 227-233.

Magurran, A.E., 1988. *Ecological diversity and its measurement*. Princeton University Press, Princeton, NJ, 179 pp.

Meyer, W.B. & Turner II, B.L., (Eds.), 1994. *Changes in land use and land cover: A global perspective.* Cambridge Univ. Press, Cambridge, UK, 537 pp.

Nault, L.R., 1997. Arthropod transmission of plant viruses: a new synthesis. *Ann. Entomol. Soc. Amer.* **90**: 521-541.

Pedigo, L.P., 1996. *Entomology and pest management*, 2nd edition. Prentice Hall, New Jersey.

Rodriguez, C.M., Madden, L.V. & Nault L.R., 1992. Diel flight periodicity of *Graminella nigrifrons* (Homoptera: Cicadellidae). *Ann. Entomol. Soc. Am.* **85**: 792-798.

Saville, N.M., Dramstad, W.E., Fry, G.L.A & Corbet, S.A., 1997. Bumblebee movement in a fragmented agricultural landscape. *Agriculture, Ecosystems and Environment* **61**: 145-154.

Schneider, J.C., 1989. Role of movement in evaluation of area-wide insect pest management tactics. *Environ. Entomol.* **18**: 868-874.

Showers, W.B, Reed, G.L., Robinson, J.F. & DeRozari, M.B., 1976. Flight and sexual activity of the European corn borer. *Environ. Entomol.* **5**: 1099-1104.

Stamps, J.A., Buechner, M. & Krishnan, V.V., 1987. The effects of edge permeability and habitat geometry on emigration from patches of habitat. *Am. Nat.* **129**: 533-552.

Stinner, R.E., Barfield, C.S., Stimac, J.L. & Dohse, L., 1983. Dispersal and movement of insect pests. *Ann. Rev. Entomol.* **28**: 319-335.

Taylor, R.A.J., Nault, L.R. & Styer, W.E., 1993. Experimental analysis of flight activity of three *Dalbulus* leafhoppers (Homoptera: Auchenorrhyncha) in relation to migration. *Ann. Entomol. Soc. Am.* **86**: 655-667.

Turchin, P., Odendall, F.J. & Rausher, M.D., 1991. Quantifying insect movement in the field. *Environ. Entomol.* **20**: 955-963.

Walker, T.J., 1978. Migration and re-migration of butterflies through north peninsular Florida: Quantification with malaise traps. *Journal of the Lepidopterists' Society* **32**: 178-190.

Westman, W.E., Strong, L.L. & Wilcox, B.A., 1989. Tropical deforestation and species endangerment: the role of remote sensing. *Landscape Ecology* **3**: 97-109.

Wiens, J.A., Schooley, R.L. & Weeks, Jr., R.D., 1997. Patchy landscapes and animal movements: do beetles percolate? *Oikos* **78**: 257-264.

Wiens, J.A., Stenseth, N.C., Van Horne, B. & Ims, R.A., 1993. Ecological mechanisms and landscape ecology. *Oikos* **66**: 369-380.

Williams, M., 1994. Forest and tree cover. In: Meyer, W.B. & Turner II, B.L.(Eds.), *Changes in Land Use and Land Cover: A Global Perspective.* Cambridge University Press, pp. 97-124.

With, K.A., Gardner, R.H. & Turner, M.G., 1997. Landscape connectivity and population distributions in heterogeneous environments. *Oikos* **78**: 151-169.

Wratten, S.D. & van Emden, H.F., 1995. Habitat management for enhanced activity of natural enemies of insect pests. In: Glen, D.M., Greaves, M.P. & Anderson, H.M. (Eds.), *Ecology and Integrated Farming Systems.* John Wiley & Sons Ltd., NY, pp. 117-145.

CHAPTER 10

LANDSCAPE MANAGEMENT AND RESIDENT GENERALIST PREDATORS IN ANNUAL CROP SYSTEMS

RICCARDO BOMMARCO *and* BARBARA EKBOM
Department of Entomology, Swedish University of Agricultural Sciences, Uppsala, Sweden

10.1 Introduction

Pest suppression using biological control instead of chemicals, is an important step towards sustainable food production. A long term objective for efficient biocontrol is to enhance abundance or persistence of resident natural enemy populations (Altieri, 1987). It is widely accepted that the large homogenous habitat represented by modern agricultural crop fields, coupled with the use of pesticides, is detrimental to natural enemies of insect pests in particular and biodiversity in general. Several empirical studies have demonstrated that a simplification of the agricultural landscape leads to a decrease in natural enemy activity, reproductive success and abundance (Flaherty, 1969; Dempster & Coaker, 1974; Landis & Haas, 1992; Corbett & Rosenheim, 1996; Bommarco, 1998a), but in general we know little about the processes creating these patterns. Interactions between insect pests, their host plants and natural enemies have most often been studied within crop fields. However, agroecosystems are much more than simply the fields of our studied crops. Adjacent crops and non-crop elements can have a substantial impact on insect communities in an agricultural setting. We can, to a certain extent, manipulate the agricultural landscape. It is possible to change the proportion of different habitats (the composition) and their arrangement (the configuration) in an agroecosystem. But at present we need more knowledge about the consequences of such changes on the population dynamics of predators and other resident species. Important questions for the future are whether it is feasible to change cultivation practices such that natural enemy efficacy is increased, and how large these alterations need to be in order to generate tangible results.

To manage agricultural landscapes efficiently we need to explore the processes that determine population growth and dispersal of predators. A fruitful way to address these issues is conceiving how reproduction and mortality of individuals and population growth of mobile generalist predators are linked to landscape composition. In particular we believe the key to answering these questions lies in understanding 1) which demographic parameters may have a large impact on predator population growth rate, 2) the contributions of different habitat types to important demographic parameters of the predators, 3) the ability of predators to move within and between different habitats in the agricultural ecosystem, and 4) how the previous points mediate the effect that spatial arrangement and edge properties of habitats have on long term population dynamics of the predator

B. Ekbom, M. Irwin and Y. Robert (eds.), Interchanges of Insects, 169-182

and suppression of the prey. In this chapter we will address the first three points presenting empirical data from a cereal agroecosystem in Sweden. Modeling predator dynamics in a spatially explicit setting is proposed as a tool to address the fourth point. The topics are discussed in the context of a generalist predatory carabid, *Pterostichus cupreus* L. (Coleoptera, Carabidae), commonly found in agroecosystems throughout Europe.

10.2 Efficacy of Generalist Predators

In a study Murdoch, Chesson, and Chesson (1985) put forward that the efficacy of generalist predators as natural enemies to agricultural pests was underestimated. They proposed that the "sit and wait" strategy of generalist predators may be efficient in suppressing prey although the predators do not possess the short generation time needed to track the pest population in a density dependent manner. The polyphagous strategy allows them to persist in an area when the prey is not present. Generalist predators have indeed demonstrated to be efficient control agents (Murdoch *et al.*, 1985; Chiverton, 1986; Riechert & Bishop, 1990; Settle *et al.*, 1996). However, the long life time, low population growth rates and comparably low dispersal capacity of many generalist predators, mean they have to survive year round in a limited area. Important prerequisites to their population persistence and efficacy as biocontrol agents are therefore that alternative food sources are accessible when the pest is not present, and that the habitat structure will allow them to redistribute to the pest outbreak areas.

The temperate zone agroecosystem is dominated by annual crops. This implies that when the annual crop fields are fallow, inhabitant generalist predators are confined to adjacent perennial habitats or perennial crops to find alternative food sources and shelter. Fortunately, several resident predators seem adapted to these conditions, and can, by moving between perennial and annual habitats, take advantage of the ephemeral but ample resource that insect pests in annual crops constitute. The seasonal invasion of annually disturbed habitats may actually be a result of evolutionary adaptations to the variable nature of agricultural ecosystem during thousands of years of cultivation (Settle *et al.*, 1996; Wissinger, 1997). Poorly managed landscapes, with respect to natural enemy persistence and efficacy, may therefore be landscapes with little access to alternative habitats. This may not only have adverse populations effects because of poor access to alternate food sources, but could also lead to increased mortality by predation or abiotic factors. Spatial barriers in the landscape limit dispersal reducing the ability of the predator to track prey in space (Kareiva, 1987), and may reduce the possibility to reach alternative habitats early and late in the growing season. Access to refuges and ability to quickly redistribute in the landscape are also critical factors for survival of predators in agroecosystems where pesticides are applied (Sherratt & Jepson, 1993).

During the past two decades a number of studies have demonstrated the importance of polyphagous natural enemies as predators of cereal aphids. Some ground beetles have been identified as potential biocontrol agents of agricultural pests (Chiverton, 1988; Chiverton and Sotherton, 1991; Wratten & Powell, 1991; Ekbom *et al.*, 1992). Studies from Sweden have demonstrated that polyphagous predators are indeed important as biological control agents of the bird-cherry oat aphid (*Rhopalosiphum padi* L.) (*e.g.*, Chiverton, 1986).

However, the success of polyphagous predators in controlling aphid outbreaks in cereals will depend on the presence and abundance of natural enemies. For this reason much research has focused on methods for enhancing natural enemy populations in cereals. Among the ideas suggested have been beetle banks and conservation headlands (see Dennis *et al.* in this book) primarily to give natural enemies better overwintering sites (Sotherton, 1985). The implicit assumptions in these suggestions have been that natural enemies will be able to access these areas, *i.e.* movement capacity is adequate, and that the landscape structure actually does have an impact on natural enemy survival, reproductive success and abundance. To better understand the effect that adjacent perennial habitats have on generalist predatory arthropods, a series of studies was performed on a selected model predator, *P. cupreus*. In this chapter we present the results of these studies.

P. cupreus is a one centimeter long, dark metallic copper-green carabid that favors open fields (Wallin, 1986) and prefers to move by walking. It can be found in a variety of open field habitats: shouldering through the dense vegetation of a ley, traversing a newly sown cereal field, or running along the ridges and furrows of an earthed up potato field. The chance of catching prey, of finding resources needed to generate offspring and risk of being killed, are particular for each such environment. In other words, the survival and reproductive success of this beetle depends on the quantity and quality of resources encountered in the different habitat types as it moves across the landscape.

10.3 Stage Specific Effects of Food Limitation

Food shortage can be an important limiting factor for the survival and reproduction of organisms in nature, ultimately affecting the growth of the population (Lenski, 1982). Individuals from several groups of generalist arthropod predators, such as carabid and cicindelid beetles (Pearson & Knisley, 1985; Sota, 1985; Juliano, 1986), web building spiders (Wise, 1983) and praying mantids (Eisenberg *et al.*, 1981; Hurd & Eisenberg, 1984), are limited by food in the field. An individual is food limited if acquired energy and nutrients through feeding are so low that reproduction or survival are reduced (*e.g.*, Pearson & Knisley, 1985). A possible cause of food shortage in nature is exploitative competition for food (Juliano & Lawton, 1990), which has also been hypothesized as a regulating mechanism in predator populations (Hairston *et al.*, 1960). Both intraspecific (Fagan & Hurd, 1994) and interspecific competition (Lenski, 1982; Niemelä, 1993) among predatory insects have been observed in field experiments.

Impelled by the evidence in the literature of the strong impact that food has on many generalist predators, our attention was directed to the study of effects of food on survival and reproduction of *P. cupreus*. A complicating fact is, however, that *P. cupreus* has several stages in its life cycle. The adult oviposits in June. Larvae develop during the summer and pupate in late summer. The young adult (teneral) ecloses in early autumn, and after a winter in diapause it reproduces the following spring. Adults of *P. cupreus* may live for two years or more (Wallin, 1985, 1986). This means that an individual beetle may occupy several different ecological niches during its life. Some stages in the life cycle of *P. cupreus* may be more sensitive to effects of food limitation than others.

Effects of food availability on reproduction and survival on different stages of *P. cupreus* were examined in feeding experiments (Bommarco, 1998b). Adult energy storage, fecundity, winter mortality, egg size, larval growth rate, larval mortality and pupal weight was varied by giving adults and larvae of *P. cupreus* different levels of food in the laboratory (Bommarco, 1998b). These feeding experiments were designed to by maximize variation of stage specific fitness factors making it possible to identify sensitivity to food shortage in the life cycle.

All stages in the life cycle of *P. cupreus* were affected by food level, directly or indirectly. Adult fecundity and body weight, however, increased markedly with feeding rate, and were the traits on which food limitation had the greatest impact (Bommarco, 1998b). In insects, most energy for reproduction comes either from larval or adult feeding (Slansky & Scriber, 1985). Larval conditions influence adult body size (Nelemans, 1988; Ernsting *et al.*, 1992; van Dijk, 1994) which in many insects is correlated to reproductive capacity (Roff, 1992). However, for *P. cupreus,* adult feeding provides most resources to reproduction. Body size did not explain variation in egg production, although a range of body sizes was represented in the experiment (Bommarco, 1998b). Restricted food availability for reproductive adults in nature may limit population growth, not only because most resources for reproduction are collected by adults, but also because of a long life span. *P. cupreus* have the opportunity for reproduction over two seasons or more (Wallin, 1985, 1986).

10.4 Agricultural Ecosystems in Sweden

In Sweden 43% of the cultivated area are cereals, 9% are other annuals, 36% are ley and other perennial habitats such as pastures, and 12% is fallow. Ley is harvested for hay or ensilage 2 to 3 times per year. Not all farms plant these crops. Agricultural cropping systems provide a setting where landscape structure is variable and the type of landscape is related to farming practices. Farms with animal production grow fodder and consequently alternate perennial and annual crops. These farms often have small fields, with a mix of crops, and have therefore an agricultural landscape with high spatial complexity. At the other extreme of the spectrum are conventional farms with large monocultured fields, a system with low spatial complexity. Furthermore, a crop rotation mainly of annual crops within conventional farming leads to high temporal variation. These differing cultivation practices can lead to contrasting resource availability among habitats, viewed at the scale of a ground beetle. Different kinds of agroecosystems provide an opportunity to explore the impact of landscape heterogeneity on predatory beetles.

10.5 Landscape Complexity and Food Limitation

The structural heterogeneity of a landscape is likely to influence the lifetime fitness of invertebrates, including mobile predatory insects. As they move across the landscape they encounter a variety of habitats that differ in resource availability, microclimate and shelter. Ground beetles such as *P. cupreus* are likely to be sensitive to landscape heterogeneity at the farm level. High rates of activity (Wallin & Ekbom, 1994), suggest that individual beetles can cross several fields in a life-time. A landscape size within the range of 2 to 50 hectares is likely to affect a population of beetles (Baars, 1979; Firle *et al.*, 1998). For a ground beetle with preference for open land (Wallin, 1986), agricultural fields with a mixture of crops constitute a variety of habitat qualities. Vital rates such as adult feeding rate, and fecundity may be influenced by the composition of this "undivided heterogeneous environment" (Addicott *et al.*, 1987).

Questions of interest are if *P. cupreus* suffers food shortage in nature, and if sensitive vital rates increase or decrease in relation to landscape complexity, measured at an appropriate scale. As an initial step towards understanding these relationships, fecundity and feeding rate (energy storage) of field collected *P. cupreus* were correlated to landscape structure measured at the scale of the life time range of *P. cupreus* (Bommarco, 1998a). Beetles were captured live from five localities, selected to encompass variation in cultivation practice and landscape structure. Landscape in these localities was characterized by a number of measures summarizing the degree of landscape complexity within the range of *P. cupreus* (Table 1).

Note that in the selected localities *P. cupreus* was likely to encounter only 2 different fields in $\text{Lövsta}_{(Conv)}$, 10 in $\text{Kasby}_{(Conv)}$, and roughly 14 different fields in the three organically farmed areas (Table 1). These are strikingly different levels of landscape heterogeneity for an insect to encounter. To measure feeding rate of captured beetles the Energy Reserve Index (ERI) was used. This index estimates energy reserves from measures of body weight and length, and was calibrated to *P. cupreus* specifics in laboratory feeding experiments with beetles given different levels of food (Bommarco, 1998b). The ERI of field collected beetles was compared to the ERI of laboratory beetles reared at *ad libitum* food levels. Field beetles are food limited in nature when they have a lower energy reserve level than *ad libitum* laboratory beetles.

Table 1. Landscape characteristics of examined conventional (Conv) or organic (Org) farms in the area of Uppsala, Sweden (59°51'N, 17°41'E). Arable fields within a 50 ha circle around the trapping sites are included in the analysis. In 1995 no trappings were performed in Solhem and Finsta (from Bommarco 1998a).

	Locality					
	LKil (Org)	Solhem (Org)	Finsta (Org)	Kasby (Conv)	Lövsta (Conv)	p *
Number of P. cupreus caught in 95/96	102/215	-/148	-/252	89/181	66/231	
Number of fields	9	19	14	10	2	
Percentage annual crops	22	32	43	80	100	
Length of cultivated perimeter (m)	2900	2700	3900	3300	700	
Mean field area (ha)	2.9(0.4)	3.2(0.7)	2.9(0.6)	5.9(1.8)	22.7(1.2)	0.03
Perimeter/area ratio (m/ha)	367(41)	311(26)	348(51)	224(27)	165(11)	0.03
Cultivated perimeter/area ratio (m/ha)	223(41)	117(15)	260(40)	136(19)	30(16)	0.001

* Statistical difference analyzed using Kruskal-Wallis test. Standard errors in parenthesis.

Field collected adults from all localities were more or less food limited (Bommarco, 1998a). Beetles from organically cultivated farms with smaller fields and a high percentage of perennial crops had as much as a 42% higher ERI, had larger body sizes, and almost three times higher fecundity than beetles from localities with low spatial complexity and large proportion annual crops (Fig. 1) (Bommarco, 1998a). For this generalist predator, fecundity appeared to increase with the degree of landscape heterogeneity and proportion perennial crops within its range of travel. The combined effect of landscape structure, crop composition and cultivation practice seems to affect important vital rates of this species.

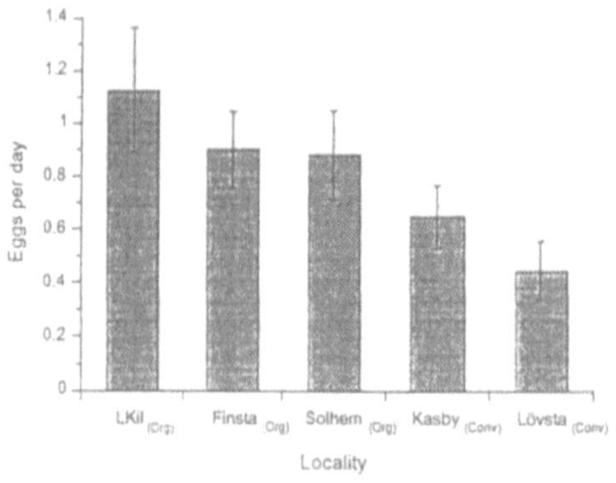

Figure 1. Egg production of *P. cupreus* in the field in different localities. Bars with standard errors (from Bommarco 1998b).

10.6 Feeding and Fecundity in a Perennial and an Annual Crop

Food limitation in nature can have two possible explanations. Predators may suffer food shortage because they compete for food, in that way depleting prey (Juliano & Lawton, 1990). An alternative explanation is that predators starve simply because the landscape they inhabit mainly consists of habitats where food is inherently difficult to find or utilize (Andrewartha & Birch, 1954). These mechanisms were examined in two habitats typical for the Swedish agricultural landscape, perennial ley and annual barley (Bommarco, 1999). The aims were to examine if *P. cupreus* is food limited in each of these habitats and to investigate if reductions in feeding rate and fecundity are caused by intraspecific competition. The design of the field experiment also made it possible to assess the actual prey availability and impact of *P. cupreus* on arthropod community composition. Enclosed plots were established in each crop by surrounding 2.5 x 2.5 m areas with plastic barriers. Individually marked *P. cupreus* were released into the plots at three densities, a control with no beetles, low density with 12 beetles/plot and high density with 62 beetles/plot. After 15 days, feeding rate, fat storage and egg load of recaptured *P. cupreus* were measured and samples were taken to assess arthropod community composition in each plot. This experiment was performed twice, early and late in the growing season 1996 (Bommarco, 1999).

Total arthropod abundance and diversity were markedly lower in barley (Fig. 2), and prey availability to *P. cupreus* was therefore lower in barley, especially early in the season. This was reflected in eggload (Fig. 3), ERI, amount of stored fat (Fig. 4) and live body weight of recaptured *P. cupreus*, which were all substantially lower in barley, indicating low feeding rates. Fat storage was to some extent reduced by intraspecific competition, but the main difference in fat storage between crops (Fig. 4) was a result of difficulty to find food in a habitat with low prey availability. Manipulating *P. cupreus* densities did not result in any changes in arthropod community composition. Similar manipulations with single spider species have resulted in weak effects (Riechert & Lockley, 1984), but an assembly of mobile generalist predators has shown pervasive effects on the prey community (Chiverton, 1986; Riechert & Bishop, 1990; Settle *et al.*, 1996). It may be necessary to consider the assemblage of generalist predators in the system to appreciate their role. These results indicate that to *P. cupreus*, barley is a habitat poor in food. This provides an explanation for the lower feeding rate and fecundity found in agricultural landscapes dominated by annual crops (Bommarco, 1998a).

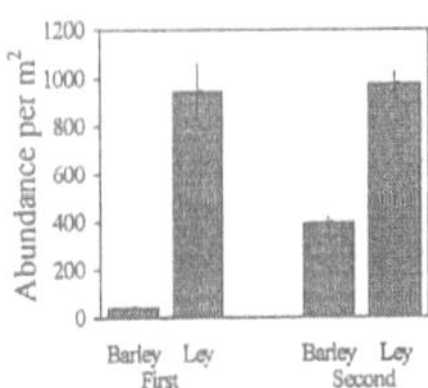

Figure 2. Total arthropod abundance per m² with standard errors in the early season and late season experiment.

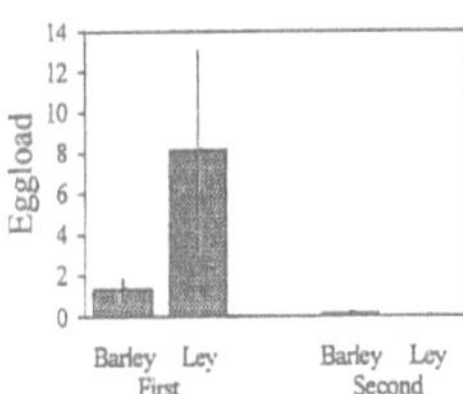

Figure 3. Mean eggload per *P. cupreus* female with standard errors in recaptured dissected beetles in the early season and late season experiment. Females carried no eggs in the second experiment in ley.

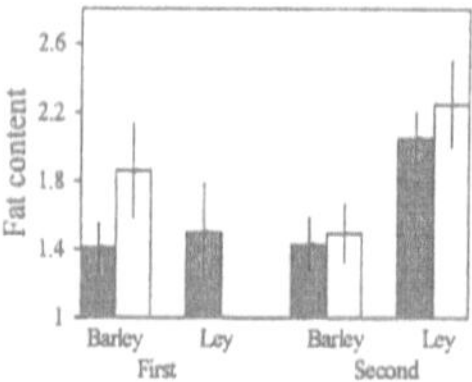

Figure 4. Fat content with standard errors in recaptured dissected beetles of *P. cupreus* at the end of each experiment, and in each crop at high (gray bars) and low (white bars) densities of *P. cupreus*. Fat content was measured by eye in dissected beetles, on a discrete scale from 1 (no fat bodies) to 3 (maximum amount of contained fat.).

10.7 Exchange of Predators Between an Annual and a Perennial Crop

In the previous sections we presented data showing that *P. cupreus* is food limited in nature and that the degree of food limitation is related to agricultural landscape complexity. Availability of perennial grasslands seems to be of particular importance to the foraging success and reproduction of these beetles, and in a lifetime an individual *P. cupreus* is likely to come across a variety of habitat types (Firle *et al.*, 1998). We have, however, little knowledge about the exchange of the beetles among different habitat types, and the willingness of these beetles to move into the fields cultivated with annual crops. The dispersal of ground beetles may be hampered at the boundaries between the habitats. The ability to cross an agricultural landscape consisting of several fields may for instance be impeded by roads, ditches and other physical barriers, or because of an innate reluctance of the organism to leave or to enter certain crops. The degree of boundary hardness between the cultivated fields may also vary over the season. As the crop grows, microclimate and food availability changes, leading to new responses at the edges, of immigrating and emigrating individuals.

Permeability is a quantitative measure of the proportion of dispersing individuals that after reaching a boundary then cross over it (Stamps *et al.*, 1987). In this section we present a field study performed at the organically managed Finsta farm 15 km west of Uppsala where permeability was estimated for *P. cupreus* dispersing between an annual cereal crop and perennial ley. We used a setup of 16 directional traps along the edge between perennial ley and spring sown oats (Fig. 5). The traps were emptied each morning and evening for four weeks starting on 4 June 1996. The directional traps made it possible to assess the number of beetles moving out from ley into oats and vice versa, during this period of time. After counting the captured beetles, they were released along the crop edge at least 50 m from the traps. Crop heights were, at the start, in ley 6 cm and oat seedlings had just emerged, and at the end crop height in ley was 71 cm and in oats 34 cm.

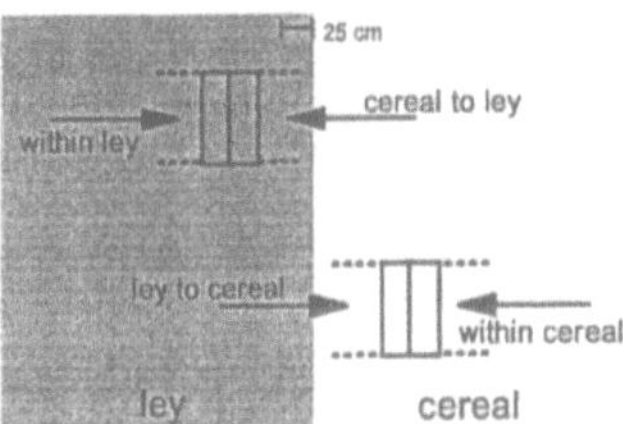

Figure 5. Experimental setup for estimating permeability. Boxes with solid lines represent traps. Each trap had a divider in the middle so that the direction from which beetles came was evident. Dotted lines represent barriers designed to channel beetles into traps. Permeability for ley is the ratio of number of beetles moving from ley to cereal divided by within ley movement. The ratio of number of beetles moving from cereal to ley divided by within cereal numbers is cereal permeability.

The results show that *P. cupreus* readily moves into the oat field even early in the season when the ground is still bare (Fig. 6). Permeabilities are in most cases near one. The first four estimates are the most reliable because the total number of beetles caught decreased the last two weeks. No significant differences were detected, but the permeability tended to drop from week one to week two for beetles moving from oats to ley. This may be caused by the fact that the oat field was practically bare during the first week presenting very little food and protection for the beetles. The low permeability for beetles moving from oats into ley in week two indicates a preference for beetles to stay in the oats during this period.

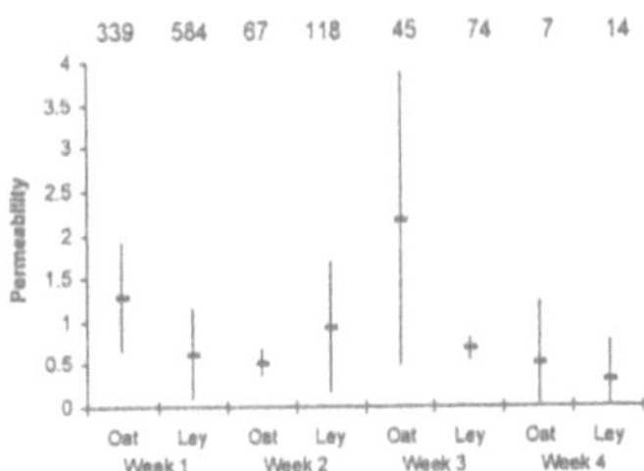

Figure 6. Mean and SD of daily permeabilites for *P. cupreus* beetles moving between an oat and a ley field. Oat denotes the permeability for beetles moving from oat to ley. Ley denotes the permeability for beetles moving from ley to oat. moving from oat to ley (noted as oat in the figure) and from ley to oat (noted as ley). Over each bar the total number of beetles caught each week is noted.

10.8 Modeling Beetle Populations in Contrived Agricultural Landscapes

The ultimate goal in landscape planning for better natural enemy control of pest populations is enhancement of resident predator populations. The difficulty in experimentally manipulating the landscape to study long term effects on predator and pest populations is obvious. Amalgamating our empirical knowledge with theoretical modeling provides a way to identify and explore crucial ecological processes in the landscape. Modeling studies have shown that having high population of natural enemies at the time of pest colonization is a combination that will result in adequate suppression of pest populations (Ekbom *et al.*, 1992; Ives & Settle, 1997). It is therefore essential to determine what composition and configuration of the landscape will encourage population growth of generalist predators.

Our experience of generalist predator dispersal, interactions with prey and reproductive limitations lead to certain considerations we believe important when modeling the system. Because many generalist predators move by walking their movement may often be approximated with diffusion and can be considered limited (Wetzler & Risch, 1984; Fagan, 1997; Firle *et al.*, 1998). Such limitations in dispersal together with strong predator-prey interactions have important implications for spatial dynamics of predators and prey even in homogenous environments (Holmes *et al.*, 1994; Maron & Harrison, 1997). Furthermore, if the generalist predator is limited by food, prey must be accessible. These limitations imply that it is important that patches of prey are within the movement range of the predators, and that the spatial location and arrangement of prey patches can greatly influence population dynamics.

These considerations lead us to the conclusion that in order to answer questions about landscape dynamics in this system, we need to apply spatially explicit models (Dunning *et al.*, 1995), where habitat quality may be described in terms of prey abundance. In the model energetic gains from prey consumption should be weighed against costs of locating food. Sparse prey availability will cause the beetles to range over wide areas looking for prey and depleting stored energy. Moving between fields with different crops does not seem to be a problem as shown by the permeability results. However, if distances are long then beetles may encounter impediments more difficult to traverse than field edges (Mauremooto *et al.*, 1995) and permeability may become an issue.

We have developed an individual based model in which movement parameters are estimated from field data on movement behavior (Firle *et al.*, 1998). By adding information on habitat specific demographic rates it will be possible to study not only movement capacity but also reproductive and energetic parameters. We will be able to determine whether or not the beetle's demand for sufficient food for survival, growth and reproduction is met in any particular landscape design. An apparent weakness in this approach is the necessity for detailed data from which to estimate parameters. We may, however, once we know it is possible, exclude certain unnecessary parameters and simplify the model. It is, for instance, possible to substitute detailed movement behavioral rules with simple diffusion to study dispersal of carabid beetles at large spatial scales (Firle *et al.*, 1998).

10.9 Conclusion

The aim of this chapter was to reach a better understanding about how realized reproduction and survival of a mobile organism is linked to landscape composition. Among the multitude of ecological interactions that affect reproduction and survival of an organism, we have attempted to pinpoint factors of critical importance to individuals and to population growth of a generalist predator. The results presented suggest that food limitation for adult long lived predatory carabids is an important limiting factor for population growth. The extent to which food limitation occurs is correlated with the management practice and the resulting complexity of the agricultural landscape. In a landscape consisting mainly of annual crops, as is often the case in Swedish agriculture, access to perennial habitats such as ley or grassy field margins seem to be vital to the feeding and reproduction of adult *P. cupreus* (Bommarco, 1998a,a, b), and is therefore probably also important to population growth and dynamics. This situation may be a reality, not only for *P. cupreus*, but for a range of predatory carabids that migrate between annual and perennial systems (Wallin, 1985; Thomas *et al.*, 1991). A variety of habitats within range means access to important sources of alternate food, in addition to refuge and overwintering sites (Corbett & Rosenheim, 1996). The results presented in this chapter suggest the removal of critical food sources as a mechanism for a the decrease in natural enemy activity and abundance often observed in simplified the agricultural landscapes.

Some ground beetles have been identified as potential biocontrol agents of agricultural pests (Chiverton, 1988; Chiverton & Sotherton, 1991; Wratten & Powell, 1991; Ekbom *et al.*, 1992). *P. cupreus* is a representative of this group of generalist predators that preys on invertebrates including aphids in cereal fields, and have the potential to reduce cereal aphid numbers (Chiverton, 1987, 1988). Adults of *P. cupreus* are likely to range over several fields during a life time (Firle *et al.*, 1998). Barley and ley mean widely different habitat qualities for *P. cupreus* (Bommarco, 1999), but it shows no reluctance for moving into barley, even early in the season (Fig. 6), despite the risk of experiencing extreme food limitation there (Bommarco, 1999). This is positive news to the pest manager, who wants natural enemies to be in place early in the season when immigrating pests first establish (Ekbom *et al.*, 1992; Ives & Settle, 1997).

The question remains, however, about the optimal spatial arrangement and proportion of perennial and annual habitats in the landscape, that would yield efficient long term natural biocontrol. We believe a fruitful approach for understanding this issue is to model the system in spatially explicit models. These model could then be used to identify important ecological processes and to understand in what landscape biocontrol by resident natural enemies may be effective. Gathering empirical evidence should then be more efficient when guided by the model framework.

Acknowledgments

We acknowledge financial support for this research from the Swedish Council for Forestry and Agricultural Research (SJFR), Carl Tryggers Stiftelse för Vetenskaplig Forskning and Oscar and Lili Lamm's foundation.

10.10 References

Addicott J.F., Aho, J.M., Antolin, M.F., Padilla, D.K., Richardsson, J.S. & Soluk, D.A., 1987. Ecological neighborhoods: scaling environmental patterns. *Oikos* **49**:340-346.

Altieri, M.A., 1987. *Agroecology the scientific basis of alternative agriculture*. 2. (Ed.) Westview Press. Boulder. USA.

Andrewatha, H.G. & Birch, L.C., 1954. *The distribution and abundance of animals*. University of Chicago Press. Chicago. USA.

Baars, M.A., 1979. Patterns of movement of radioactive Carabid beetles. *Oecologia* **44**:125-140.

Bommarco, R., 1998a. Reproduction & energy reserves of a predatory carabid beetle relative to agroecosystem complexity. *Ecological Applications* **8**:846-853.

Bommarco, R., 1998b. Stage sensitivity to food limitation of a generalist arthropod predator. *Environmental Entomology* **27**:863-869.

Bommarco, R., 1999. Feeding reproduction and community impact of a carabid predator in two agricultural habitats. *Oikos* (in press).

Chiverton, P., 1986. Predator density manipulation and its effects on *Rhopalosiphum padi* (Homoptera: Aphididae) in spring barley. *Annals of Applied Biology* **109**:49-60.

Chiverton, P.A., 1987. Predation of *Rhopalosiphum padi* (Homoptera: Aphididae) by polyphagous predatory arthropods during the aphids' pre-peak period in spring barley. *Annals of Applied Biology* **111(2)**:257-269.

Chiverton, P.A., 1988. Searching behaviour and cereal aphid consumption by *Bembidion lampros* and *Pterostichus cupreus*, in relation to temperature and prey density. *Entomologia experimentalis et Applicata* **47**:173-182.

Chiverton, P.A. & Sotherton, N.W., 1991. The effect on beneficial arthropods of the exclusion of herbicides from cereal crop edges. *Journal of Applied Ecology* **28**:1027-1039.

Corbett, A. & Rosenheim, J.A., 1996. Impact of a natural enemy overwintering refuge and its interaction with the surrounding landscape. *Ecological Entomology* **21**:155-164.

Dempster, J.P. & Coaker, T.H., 1974. Diversification of crop ecosystems as a means of controlling pests. In: Jones, D.P. & Solomon, M.E. (Eds.), *Biology in pest and disease control*. Blackwell Scientific Publications, Oxford, UK.

Dunning, J.B., Stewart, D.J., Danielson, B.J., Noon, B.R., Root, T.L. Lambersson, R.H. & Stevens, E.E., 1995. Spatially explicit population models: current forms and future uses. *Ecological Applications* **5(1)**:3-11.

Dunning, J.B., Danielson, B.J. & Pulliam, H.R., 1992. Ecological processes that affect populations in complex landscapes. *Oikos* **65(1)**:169-175.

Eisenberg, R.M., Hurd, L.E. & Bartley, J.A., 1981. Ecological consequences of food limitation for adult mantids (*Tenodera ardifolia sinensis* Saussure). *The American Midland Naturalist* **106(2)**:209-218.

Ekbom, B.S., Wiktelius, S. & Chiverton, P.A., 1992. Can polyphagous predators control the bird cherry-oat aphid (*Rhopalosiphum padi*) in spring cereals? *Entomologia experimentalis et Applicata* **65**:215-223.

Ernsting, G., Isaaks, J.A. & Berg, M.P., 1992. Life cycle and food availability indices in *Notiophilus biguttatus* (Coleoptera, Carabidae). *Ecological Entomology* **17**:33-42.

Fagan, W.F., 1997. Introducing a "boundary-flux" approach to quantifying insect diffusion rates. *Ecology* **78(2)**:579-587.

Fagan, W.F. & Hurd, L.E., 1994. Hatch density variation of a generalist arthropod predator: population consequences and community impact. *Ecology* **75(7)**:2022-2032.

Firle, S., Bommarco, R., Ekbom, B. & Natiello, M., 1998. The influence of movement and resting behavior on the range of three carabid beetles. *Ecology* **79(6)**:2113-2122.

Flaherty, D.L., 1969. Ecosystem trophic complexity and Willamette mite, *Eotetranychus willamettei* Ewing (Acarina: Tetranychidae), densities. *Ecology* **50**:911-915.

Hairston, N.G., Smith, F.E. & Slobodkin, L.B., 1960. Community structure, population control, and competition. *American Naturalist* **94**:421-425.

Holmes, E.E., Lewis, M.A., Banks, J.E & Veit, R.R., 1994. Partial differential equations in ecology: spatial interactions and population dynamics. *Ecology* **75(1)**:17-29.

Hurd, L.E. & Eisenberg, R.M., 1984. Experimental density manipulations of the predator *Tenodera sinensis* (Orthoptera: Mantidae) in an old field community. I. Mortality, development, and dispersal of juvenile mantids. *Journal of Animal Ecology* **53**:269-281.

Ives, A.R. & Settle, W.H., 1997. Metapopulation dynamics and pest control in agricultural systems. *American Naturalist* **149**:220-246.

Juliano, S.A., 1986. Food limitation of reproduction and survival for populations of *Brachinus* (Coleoptera: Carabidae). *Ecology* **67(4)**:1036-1045.

Juliano, S.A. & Lawton, J.H., 1990. Extrinsic vs. intrinsic food shortage and the strength of feeding links: effects of density and food availability on feeding rate of *Hyphydrus ovatus*. *Oecologia* **83**:535-540.

Kareiva, P., 1987. Habitat fragmentation and the stability of predator-prey interactions. *Nature* **326**:388-390.

Landis, D.A. & Haas, M.J., 1992. Influence of landscape structure on abundance and within-field distribution of European corn borer (Lepidoptera: Pyralidae) larval parasitoids in Michigan. *Environmental Entomology* **21**:409-416.

Lenski, R.E., 1982. Effects of forest cutting on two *Carabus* species: evidence for competition for food. *Ecology* **63(5)**:1211-1217.

Maron, J.L. & Harrison, S., 1997. Spatial pattern formation in an insect host-parasitoid system. *Science* **278**:1619-1621.

Mauremooto, J.R., Wratten, S.D., Worner, S.P. & Fry, G.L.A., 1994. Permeability of hedgerows to predatory carabid beetles. *Agriculture, Ecosystems & Environment* **52(2-3)**:141-148

Murdoch, W.W., Chesson, J. & Chesson, P.L., 1985. Biological control in theory and practice. *American Naturalist* **125(3)**:344-366.

Nelemans, M.N.E., 1988. Surface activity and growth of larvae of *Nebria brevicollis* (F.) (Coleoptera, Carabidae). *Netherlands Journal of Zoology* **38(1)**:74-95.

Niemelä, J., 1993. Interspecific competition in ground-beetle assemblages (Carabidae): what have we learned? *Oikos* **66**:325-335.

Pearson, D.L. & Knisley, C.B., 1985. Evidence for food as limiting resource in the life cycle of tiger beetles (Coleoptera: Cicinelidae). *Oikos* **45**:161-168.

Riechert, S.E. & Bishop, L., 1990. Prey control by an assemblage of generalist predators: spiders in garden test systems. *Ecology* **71**:1441-1450.

Riechert, S.E. & Lockley, T., 1984. Spiders as biological control agents. *Annual Review of Entomology*.**29**:299-320.

Roff, D.A., 1992. *The evolution of life histories - theory and analysis*. Chapman & Hall, New York, USA.

Settle, W.H., Ariawan, H., Astuti, E.T., Cahyana, W, Hakim, A.L., Hindayana, D, Lestari, A.S., Pajarningish & Sartanto, 1996. Managing tropical rice pests through conservation of generalist natural enemies and alternative prey. *Ecology* **77(7)**:1975-1988.

Sherratt, T.N. & Jepson, P.C., 1993. A metapopulation approach to modelling the long-term impact of pesticides on invertebrates. *Journal of Applied Ecology* **30**:696-705.

Slansky, F. Jr. & Scriber, J.M., 1985. Food consumption and utilization. In: Kerku, G.A. & Gilbert, L.I (Eds.), *Comprehensive insect physiology, biochemistry and pharmacology*, vol. 4, pp. 87-163. Pergamon, Oxford, UK.

Sota, T., 1985. Limitation of reproduction by feeding condition in a carabid beetle, *Carabus yaconinus*. *Researches on Population Ecology* **27**:171-184.

Sotherton, N.W., 1985 The distribution and abundance of predatory Coleoptera overwintering in field boundaries. *Annals of applied biology* **106**:17-21.

Stamps, J.A., Buechner, M. & Krishnan, V.V., 1987. The effects of edge permeability and habitat geometry on emigration from patches of habitat. *American Naturalist* **129**:533-552.

Thomas, M.B., Wratten, S.D. & Sotherton, N.W., 1991. Creation of 'island' habitats in farmland to manipulate populations of beneficial arthropods: predator densities and emigration. *Journal of Applied Ecology* **28**:906-917.

van Dijk, Th. S., 1994. On the relationship between food, reproduction and survival of two carabid beetles: *Calathus melanocephalus* and *Pterostichus versicolor*. *Ecological Entomology* **19**:263-270.

Wallin, H., 1985. Spatial and temporal distribution of some abundant carabid beetles (Coleoptera: Carabidae) in cereal fields and adjacent habitats. *Pedobiologia* **28**:19-34.

Wallin, H., 1986. Habitat choice of some field-inhabiting carabid beetles (Coleoptera: Carabidae) studied by recapture of marked individuals. *Ecological Entomology* **11**:457-466.

Wallin, H. & Ekbom, B., 1994. The influence of hunger level and prey densities on movement patterns in three species of carabid beetles. *Environmental Entomology* **23(5)**:1171-1181.

Wetzler, R.E & Risch, S.J., 1984. Experimental studies of beetle diffusion in simple and complex crop habitats. *Environmental Entomology* **53**:1-19.

Wise, D.H., 1983. Competitive mechanisms in a food-limited species: relative importance of interference and exploitative interactions among labyrinth spiders (Araneae: Araneidae). *Oecologia* **58**:1-9.

Wissinger, S.A., 1997. Cyclic colonization in predictably ephemeral habitats: a template for biological control in annual crop systems. *Biological Control* **10**:4-15.

Wratten, S.D. & Powell, W., 1991. Cereal aphids and their natural enemies. In: Firbank, L.G., Carter, N., Darbyshire, J.F. & Potts, G.R.(Eds.), *The ecology of temperate cereal fields*. The 32nd symposium of the British ecological society. Blackwell Scientific publications, London. UK. pp 233-258.

CHAPTER 11

PARASITOID COMMUNITY STRUCTURE

Implications for Biological Control in Agricultural Landscapes

PAUL C. MARINO
Department of Biology, University of Charleston, Charleston, SC

DOUGLAS A. LANDIS
Department of Entomology and Center for Integrated Plant Systems, Michigan State University, E. Lansing, MI

11.1 Introduction

Modern agricultural landscapes range in structure from the highly simplified containing little in the way of non-crop habitat, to the very complex, containing small patches of cropland nested within extensive and interconnected areas of non-crop vegetation. This diversity in the structure of agricultural landscapes can have strong impacts on the diversity, abundance and effectiveness of insect natural enemies that occur within crops (Szentkriralyi & Kozar, 1991; Kruess & Tscharntke, 1994). Understanding how these landscape features affect the interactions between crops, pests and their natural enemies is a complex problem that can significantly impact the success or failure of insect biological control (Landis, 1994).

Parasitoids are particularly important natural enemies because of their great diversity and effectiveness as agents of biological control (Hassell, 1986, LaSalle, 1993). Parasitoid effectiveness can often be enhanced by habitat manipulation either within or outside of fields (Powell, 1986). Extra-field vegetation in agroecosystems can influence parasitoid population and community dynamics either directly, by providing food and shelter, or indirectly, through the diversity and abundance of primary and alternate hosts (Altieri *et al.*, 1983).

In this chapter we address how these direct and indirect effects, mediated by the structure, diversity, extent and successional stage of extra-field vegetation, can influence parasitoid population and community structure. We discuss the physical characteristics of agricultural landscapes that distinguish them from primeval landscapes, the resource needs of parasitoids, and the importance of extra-field habitats in providing these needs. We also provide evidence suggesting that agricultural landscape structure can impact the effectiveness of parasitoids as agents of biological control. While we focus our discussion on north temperate agroecosystems and mainly use examples from our work in the North Central U.S., these concepts should be applicable to other parts of the world as well.

B. Ekbom, M. Irwin and Y. Robert (eds.), Interchanges of Insects, 183-193

11.2 The Changing Agricultural Landscape

Most modern agroecosystems differ dramatically from primeval landscapes from which they were derived. Perhaps the most important factor mediating these changes is the extent and frequency of disturbance (Merriam, 1988; Landis & Menalled, 1998). For example, in the North Central region of the U.S., the pre-European settlement landscape was a mix of forest, savanna, prairie and wetland (Auclair, 1976). As with most ecosystems, periodic natural disturbances such as fires, treefalls or floods were common and resulted in a mosaic of successional stands within this primeval landscape. The advent of European settlement transformed the landscape. Humans converted forests and prairies into agricultural croplands principally by instituting larger and more uniform disturbances such as large-scale clearing, burning and plowing. In addition, these disturbances became vastly more intense (removing most existing plants and animals) and frequent (several times per season: *e.g.*, plowing, cultivation and harvest). The crops planted were primarily annual species grown in monoculture. As such, native and introduced herbivores and their associated parasitoids were confronted with an increasingly larger and interconnected array of highly disturbed early successional habitats interspersed with an increasingly smaller and fragmented array of mid and late successional habitats (woodlots and fencerows). These changes continue to occur as new production methods are adopted. Particularly noticeable has been the enlargement of field size at the expense of hedgerows, fencerows and woodlots that has taken place in many areas.

The transformation and fragmentation of natural ecosystems by agriculture often results in the loss of species and the disruption of food webs (Diamond & May, 1981; Wilcove *et al.*, 1986). Crop plants themselves and the weedy early and mid successional habitats characteristic of agricultural landscapes are characterized by low plant species diversity and by plants with little architectural complexity (sensu Lawton, 1983). Plants having simple architectures have fewer species of insects (pests and natural enemies) living on them than later-successional, more architecturally complex plant communities (Murdoch *et al.*, 1972; Lawton, 1978; Lawton & Schroder; 1977, Southwood *et al.*, 1979; Cornell, 1986; Hawkins & Lawton, 1987; Stinson & Brown, 1983; Brown, 1991). In contrast, agricultural landscapes retaining more of the vegetational complexity of primeval landscapes (*e.g.* those having abundant hedges, meadows, small forests and wetlands) have a greater biomass and diversity of parasitoids, than uniform agricultural landscapes having little non-arable land (Ryszkowski & Karg, 1991; Ryszkowski *et al.*, 1993). One reason for this may be that within croplands, the habitat and structural diversity provided by non-crop habitats may provide increased resources for parasitoids, as does within-field diversity (Powell, 1986). These resources may include adult food (nectar and pollen), appropriate microclimates within the relatively harsh agricultural landscape and alternate hosts necessary for improving parasitoid survival and effectiveness.

11.3 Parasitoid Resources Provided by Extra-Field Habitats

11.3.1 FLORAL RESOURCES

Many parasitoids feed as adults and use wildflowers as food (van Emden, 1963, 1965; Jervis *et al.*, 1993). These floral resources have been shown to enhance parasitoid longevity and fecundity (Leius, 1963, 1967; Foster & Ruesink, 1984; Hagley & Barber, 1992; Idris & Grafius, 1995). The presence of floral resources in extra-field habitats may impact the distribution, abundance and diversity of parasitoids. In Michigan, the Ichneumonid *Eriborus terebrans* (Gravenhorst), a specialist attacking the European corn borer (*Ostrinia nubilalis* Hübner), lives longer when allowed to feed on flowers (Landis & Marino, 1999) or a sugar water solution (Dyer & Landis, 1996) versus water alone. Landis & Haas (1992) found that rates of parasitism by *E. terebrans* are highest near the wooded edges of maize fields. *Eriborus terebrans* has no alternate hosts in Michigan that would account for this distribution, rather, increased rates of parasitism by *E. terebrans* near field edges was attributed in part to the increased availability of adult food resources (Dyer, 1995; Dyer & Landis, 1996, 1977a). These data suggest that like adult food resources within crop fields, agricultural landscapes that contain extra-field adult food resources can enhance the effectiveness of Hymenopteran parasitoids. However, whether this use of, and movement between, extra-field floral resources and field crops is a common phenomenon remains unclear.

11.3.2 MODERATED MICROCLIMATES

The simplification of agricultural landscapes has also created relatively harsher microclimates that may limit the abundance and diversity of parasitoids and/or restrict parasitoids to favorable microclimates found in extra-field habitats (Dyer & Landis, 1997a). Parasitoid longevity can be greatly reduced at laboratory temperatures generated to simulate the conditions found in crop fields prior to canopy closure. Longevity of *Xanthopimpla stemmator* Thunberg (Hymenoptera: Ichneumonidae) decreased from 42 d at 20°C, to 15 d at 28°C (Hailemichael & Smith, 1994), and longevity of female *Ooencyrtus papilianis* Ashmead (Hymenoptera: Eulophidae) decreased from 10.4 d at 15°C to 1.7 d at 35°C (Rahim *et al.*, 1991). Dyer and Landis (1996) comparing the longevity of caged *E. terebrans* in cornfields, woodlots, wooded fence rows, and herbaceous vegetation found greater longevity in the more moderate microclimate of woodlots vs. early season cornfields (Fig. 1). They suggested that higher abundance of *E. terebrans* near the edges of corn fields was due not only to increased adult food resources as discussed above, but was likely also a consequence of more favorable microclimates. Studies of the diurnal behavior of *E. terebrans* indicated that host search primarily occurred during morning hours, with wasps entering a state of inactive rest during the afternoon. However, wasps were significantly more active on hotter afternoons (walking and flying) apparently in an attempt to leave the stressful environment (Dyer & Landis, 1997b).

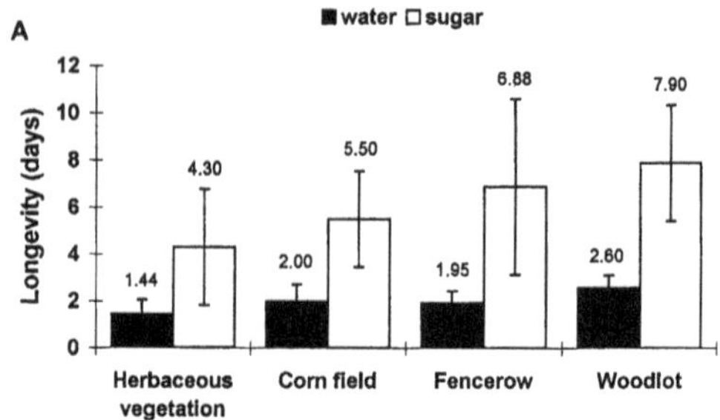

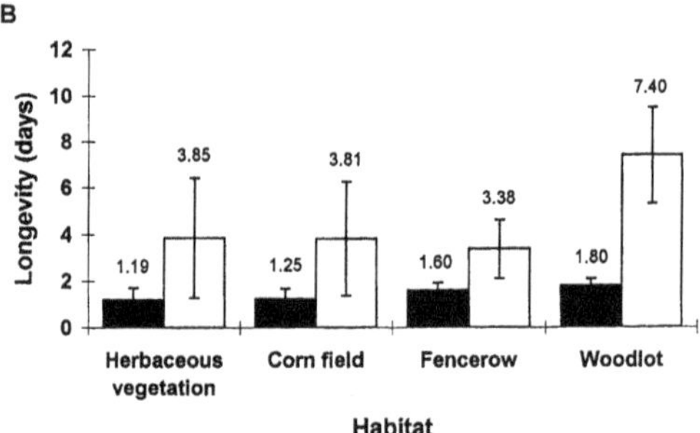

Figure 1. Early-season longevity (mean ± SE) of adult *Eriborus terebrans* confined in four habitats with either water alone, or sugar water. (A) females, (B) males. (after Dyer & Landis, 1996).

11.3.3 ALTERNATE HOSTS

Lastly, the presence or absence of alternate hosts associated with extra-field habitats can also impact parasitoid populations in agroecosystems (Doutt & Nakata, 1973; Powell, 1986). For example, the egg parasite *Anagrus epos* Girault which attacks the grape leafhopper, *Erythroneura elegantula* Osborn in California vineyards is most abundant in vineyards near riparian areas in which wild blackberry (*Rubus neura elegantula* Osborn) is abundant. Wild blackberry supports an alternate (overwintering) host of *A. epos* (Doutt & Nakata, 1973). Corbett and Rosenheim (1996) found that at both the field and landscape level, presence of habitats supporting alternate hosts increased the density of *A. epos*.

For generalist parasitoids, the presence of alternate hosts should have a positive impact on population size. However, the structure, type and extent of extra-field vegetation will determine the diversity and abundance of these alternate hosts. As such, we focus the remainder of this chapter examining the extent to which agricultural pest species are attacked by generalist parasitoids, the habitat characteristics of the alternate hosts of generalist parasitoids, and evidence that generalist parasitoids, as opposed to specialist parasitoids, can be effective agents of biological control.

11.4 Parasitoid Communities, Landscape Structure and Biological Control

Some authors have observed that specialist herbivores are attacked by relatively greater proportions of specialist parasitoids and that generalist herbivores are attacked in greater proportion by generalist parasitoids (Askew & Shaw, 1986; Price, 1991, 1994). In addition, Price (1991, 1994) predicted that in early successional habitats, specialist herbivores and thus, specialist parasitoids would predominate. However, he cautioned that the degree to which this would hold true in highly disturbed agricultural systems was uncertain. To explore the interactions of vegetation with insect pest and parasitoid community structure, Landis and Marino (unpub. data) examined the degree of specialization of Lepidopteran crop pests, their parasitoids, and the host plants used by the alternate hosts of those parasitoids in the North Central region of the U.S.

The major field crops of this region are maize, soybeans, wheat and alfalfa. Of the 43 Lepidopteran pests that commonly attack these crops, 93% of these are native to the region and 67% are polyphagous (feed on four or more plant families, Sheehan, 1991, 1994). However, they found that irrespective of herbivore host feeding range (polyphagous, intermediate or oligophagous), generalist parasitoids (attacking two or more host families, Sheehan, 1991, 1994) represent approximately 50% of the potential parasitoid community attacking these Lepidopteran species (Landis and Marino, unpub. data). These data suggest, for this particular assemblage of herbivore pests, that generalist parasitoids are potentially important agents of biological control. Because it is recognized that generalist natural enemies can contribute to the suppression of pest populations (Riechert & Bishop, 1990) it is important to understand the alternate host associations of these parasitoids and the types of vegetation with which those alternate hosts are associated. For example, if most generalist parasitoids have host associations with species that feed on late successional species, then this would suggest that late successional extra-field habitats might be important for the conservation of these generalist parasitoid populations. On the other hand, if most of these generalists attack alternate hosts occurring on early to mid successional plants then these habitats may be important for their conservation.

The types of vegetation with which the alternate hosts of the generalist parasitoids identified by Landis and Marino (unpub. data) were associated were examined by categorizing each alternate host as feeding on early (ruderals), mid (shrubs), late (trees), and mixed (trees + mid and/or early) successional species. In order to determine the potential frequency in which parasitoids searched for hosts in early vs. late successional habitats, parasitoids were then divided into those that included hosts in their food range that feed exclusively on trees or trees + shrubs versus those that do not. The results indicated that for these particular Lepidopteran pests, most of their generalist parasitoids (68%) of include hosts in their food range that feed exclusively on late successional species. Thus, it appears that many of these species must at some point search both early successional habitats (*i.e.*, crops) as well as late successional ones (*i.e.* trees) to locate these diverse hosts. Whether this is done by the same individuals, by different generations, or over years as a consequence of shifting host availability, is unclear. However, these results suggest that the role of late successional habitats in the conservation of these species should be examined.

11.5 Generalist Parasitoids and Landscape Structure

To what extent then are generalist parasitoids important regulators of the population sizes of herbivores? *i.e.*, even if one could conserve generalists, would this necessarily increase percent parasitism of crop pests? Finally, what role may landscape structure and habitat type play in these interactions?

11.5.1 AN EXAMPLE

Marino and Landis (1996) have demonstrated evidence for higher rates of parasitism by generalist parasitoids in an agricultural landscape having abundant versus little late successional vegetation. They examined rates of parasitism and parasitoid diversity associated with the true armyworm (*Pseudaletia unipuncta* Haworth) in structurally complex (crop fields embedded in a matrix of hedgerow, oldfield and woodlot) versus simple (crop fields in a matrix containing little non-crop vegetation) agricultural landscapes in central Michigan (Fig. 2). *Pseudaletia unipuncta* has 35 species of potential parasitoids in the North Central region of the U.S. However, a single generalist, the Braconid *Meteorus communis* Cresson, caused most parasitism and differences in rates of parasitism between complex and simple agricultural landscapes (13.1% vs. 2.4%).

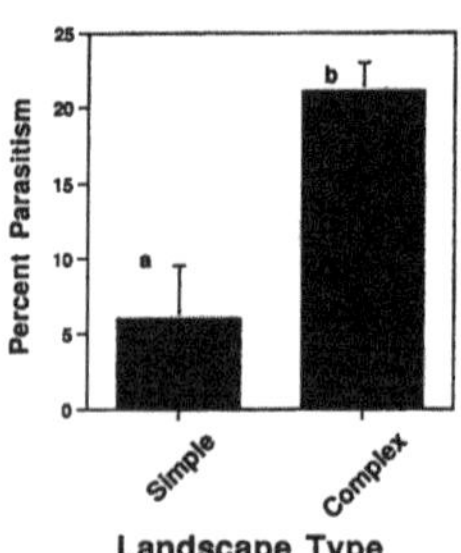

Figure 2. Percentage of *Pseudaletia unipuncta* larvae parasitized (mean ± SE) within a simple versus complex agricultural landscape in Ingham County, Michigan, USA. N = 3 fields per landscape type, $P < 0.05$ ANOVA. (after Marino & Landis, 1996).

11.5.2 ROLE OF ALTERNATE HOSTS

Marino and Landis (1996) hypothesized that alternate hosts may explain this result. All the alternate hosts of *M. communis* are exposed larvae that feed primarily on trees and shrubs common to hedgerows and woodlots in central Michigan. For example, *Prunus serotina* and *P. virginiana* are hosts for five of the seven alternate insect hosts of *M. communis*. Marino and Landis (1996) hypothesized that abundance and proximity of preferred habitats for the alternate hosts of *M. communis* may account for the observed differences in rates of parasitism between the simple vs. the complex landscape.

11.5.3 AN EVOLUTIONARY PERSPECTIVE

The association between Lepidopteran pests in the North Central U.S. and generalist Hymenopteran parasitoids having alternate hosts associated with late successional habitats, may not be surprising when one considers their evolutionary history. The primeval landscape of the region in which these herbivores and their parasitoids evolved contained abundant stable and or late successional elements. In the northern and eastern portion of the region, forest habitats predominated. In the prairies, a stable habitat in itself, forested elements were also frequently interspersed, *e.g.* along riparian corridors. This type of landscape structure favored polyphagous herbivores, which were in turn attacked by generalist parasitoids (see Price, 1991, 1994 for reasoning). As agriculture expanded into the region, these polyphagous herbivores must have occasionally encountered and readily incorporated the nutritious and poorly defended crop plants into their host ranges. Perhaps even preferring them to their native hosts. However, with this shift to early successional habitats, the potential parasitoid community still remained one dominated by generalists adapted to exploiting a variety of hosts in late successional habitats. In this light, that generalists parasitoid attacking these pests have alternate hosts associated with trees, does make sense.

In contrast, it has been suggested that many parasitoids are, in general, highly habitat specific and are more likely to attack taxonomically unrelated insects found in one habitat then they are to attack taxonomically related insects occupying different habitats (Townes, 1962, 1972; van Alphen & Vet, 1986; Altieri *et al.*, 1993). Thus, whether *M. communis'* propensity to search both forested and cultivated habitats for hosts is typical or atypical of generalist parasitoids is unclear. It is also important to note that the above scenario regarding generalist parasitoids and late successional habitats is not exclusive. For other assemblages of herbivore pests and their associated parasitoids early successional extra-field habitats may play a similarly important role.

11.5.4 CHALLENGES IN EVALUATING EFFECTIVENESS OF GENERALIST PARASITOIDS

High host-specificity has been emphasized as an important characteristic of effective natural enemies (DeBach, 1964; Huffaker & Messenger, 1976) and some have argued that specialists may be particularly important in simplified agroecosystems (Sheehan, 1986). Importation biological control efforts have often focused on specialist parasitoids and there is evidence that they establish at somewhat higher rates than polyphagous parasitoids

(Stilling, 1990). However, both mathematical models (Murdoch *et al.*, 1985) and empirical studies (Riechert & Bishop, 1990) indicate that polyphagous natural enemies can also effectively control pest populations.

In their study, Marino and Landis (1996) found that generalist parasitoids *(M. communis* and *Glyptapanteles militaris*) provided the greatest impact on *P. unipuncta*, however, the relatively low overall parasitism called into question the potential impact of these generalists in terms of biological control. This may be because these species are inherently inferior, or may be a function of the agroecosystems in which they were studied.

All agroecosystems, even those which retain a relatively complex structure, represent significant alterations of the landscape in which host and parasitoids evolved. In these landscapes it may be difficult to determine the true potential importance of generalist parasitoids. Most studies exploring parasitism of herbivore pests are performed in agroecosystems. In these landscapes, the scarcity of adult food sources, appropriate microclimates or alternate hosts, may be expected to result in reduced parasitoid populations, diversity and species richness of the parasitoid community. If so, then the potential importance of generalist parasitoids as agents of biological control may be consistently underestimated.

11.6 Summary and Conclusions

In the North Central U.S. at least, it appears that most pest Lepidoptera on the major field crops are native species that have moved onto crops from the surrounding native vegetation and many of these pests are associated with later successional habitats. As agriculture became more extensive these pests moved onto field crops. Although we know that most of these herbivores have parasitoid faunas containing many generalists, we do not know the extent to which these parasitoids move between extra-field and cultivated habitats. We do know that a number of parasitoids forage for food or hosts in both extra-field and cultivated habitats (Doutt & Nakata, 1973; Corbett & Rosenheim, 1996; Marino & Landis, 1996; Dyer & Landis, 1997). However, others have suggested that many parasitoids are habitat specific (Townes, 1972; Altieri *et al.*, 1993). If movement between extra-field and cultivated habitats occurs frequently, then the presence of extra-field habitats and their associated alternate hosts may have a significant positive impact on biological control by generalist parasitoids. Conversely, if there is little movement of generalist parasitoids from non-crop into cultivated habitats or if specialist parasitoids are responsible for most parasitism of pest species, then the presence of extra-field habitats becomes less critical to the success of biological control using parasitoid natural enemies.

Extra-field non-crop habitats can provide adult food resources, favorable microclimates and alternate hosts, all of which may enhance the abundance and diversity of parasitoids in agroecosystems. Because many parasitoids (both specialists and generalists) require these resources and because many parasitoids have generalized host ranges (Townes, 1972, Krombein, 1979) it follows that the resources necessary to support an abundant and diverse parasitoid fauna may necessitate a complex mosaic of vegetational structure. To begin to understand the extent, type and importance, of extra-field vegetational structure necessary to promote more effective biological control by parasitoids will require additional studies at the field and especially landscape scale.

Acknowledgments

We wish to thank Fabian Menalled for reading and commenting on the manuscript. The work summarized here was supported by USDA-LISA grant LWF 62-016-02942, USDA-SARE grant LWF 62-016-03508 and from the National Science Foundation Long-Term Ecological Research (LTER) program in Agricultural Ecology at Michigan State University DEB-92-1171.

11.7 References

Altieri, M.A., Cure, J.R & Garcia, M.A., 1993. The role and enhancement of parasitic Hymenoptera biodiversity in agroecosystems. In: LaSalle, J. & Gauld, I.D. (Eds.), *Hymenoptera and Biodiversity*. C.A.B. International, Oxon, pp. 257-276.

Askew, R.R. & Shaw, M.R., 1986. Parasitoid communities: their size, structure, and development. In: Waage, J. & Greathead, D. (Eds.), *Insect Parasitoids*. Academic Press, London, pp. 225-264.

Auclair, A.U., 1976. Ecological factors in the development of intensive-management ecosystems in the midwestern United States. *Ecology* **57**:431-444.

Brown, V.K., 1991. The effects of changes in habitat structure during succession in terrestrial communities. In: Bell, S., McCoy, E.D. & Mushinsky, H. (Eds.), *Habitat Structure: The Physical Arrangement of Objects in Space*. Chapman and Hall, London, pp. 141-168.

Corbett, A. & Rosenheim, J.A., 1996. Impact of a natural enemy overwintering refuge and its interaction with the surrounding landscape. *Ecological Entomology* **21**:155-164.

Cornell, H.V., 1986. Oak species attributes and host size influence cynipine wasp species richness. *Ecology* **67**:1582-1592.

Debach, P., 1964. *Biological Control of Insect Pests and Weeds*. Reinhold Publishing Corp., New York.

Diamond, J.M. & May, R.M., 1981. Island bigeography and the design of natural reserves. In: May, R.M. (Ed.), *Theoretical Ecology: Principles and Applications*. Blackwell Scientific Publications, Oxford, pp. 238-252.

Doutt, R.L. & Nakata, J., 1973. The *Rubus* leafhopper and its egg parasitoid: an endemic biotic system useful in grape pest management. *Environmental Entomology* **3**:381-386.

Dyer, L.E., 1995. Non-crop habitats and the conservation of *Eriborus terebrans* (Gravenhorst) (Hymenoptera: Ichneumonidae), a parasitoid of the European corn borer, *Ostrinia nubilalis* (Hübner) (Lepidoptera: Pyralidae). Ph.D. Dissertation, Department of Entomology, Michigan State University, E. Lansing, MI.

Dyer, L.E. & Landis, D.A., 1996. Effects of habitat, temperature, and sugar availability on longevity of *Eriborus terebrans* (Hymenoptera: Ichneumonidae). *Environmental Entomology* **25**:1192-1201.

Dyer, L.E. & Landis, D.A., 1997a. Influence of noncrop habitats on the distribution of *Eriborus terebrans* (Hymonoptera: Ichneumonidae) in cornfields. *Environmental Entomology* **26**: 924-932.

Dyer, L.E. & Landis, D.A., 1997b. Diurnal behavior of *Eriborus terabrans* (Hymenoptera: Ichneumonidae). *Environmental Entomology* **26**: 1385-1392.

Foster, M.A. & Ruesink, W.G., 1984. Influence of flowering weeds associated with reduced tillage in corn on a black cutworm (Lepidoptera: Noctuidae) parasitoid, *Meteorus rubens* (Nees von Esenbeck). *Environmental Entomology* **13**:664-668.

Hawkins, B.A. & Lawton, J.H., 1987. Species richness for parasitoids of British phytophagous insects. *Nature* **326**:788-790.

Hagley, A.C. & Barber, D.R., 1992. Effect of food sources on the longevity and fecundity of *Pholetesor ornigis*

(Weed) (Hymenoptera: Braconidae). *Canadian Entomologist* **124**:341-346.

Hailemichael, Y. & Smith, Jr., J.W., 1994. Development and longevity of *Xanthopimpla stemmator* (Hymenoptera: Ichneumonidae) at constant temperatures. *Annals of the Entomological Society of America* **87**:874-878.

Hassell, M.P., 1986. Parasitoids and population regulation. In: Waage, J. & D. Greathead, D. (Eds.), *Insect Parasitoids*. Academic Press, Orlando, pp. 201-224.

Huffaker, C.B. & Messenger, P.S., 1976. *Theory and Practice of Biological Control*. Academic Press, New York.

Idris, A.B. & Grafius, E., 1995. Wildflowers as nectar sources for *Diadegma insulare* (Hymenoptera: Ichneumonidae), a parasitoid of diamondback moth (Lepidoptera: Yponomeutidae). *Environmental Entomology* **24**:1726-1735.

Jervis, M.A., Kidd, N.A.C, Fitton, M.G., Huddleston, T. & Dawah, H.A., 1993. Flower-visiting by Hymenopteran parasitoids. *Journal of Natural History* **27**:67-105.

Krombein, K.V., Hurd, Jr.,P.D., Smith, D.R. & Burks, B.D., 1979. Catalog of Hymenoptera in America north of Mexico. Smithsonian Institution Press, Washington, DC.

Kruess, A. & Tscharntke, T., 1994. Habitat fragmentation, species loss and biological control. *Science* **264**:1581-1584.

LaSalle, J., 1993. Parasitic Hymenoptera, biological control and biodiversity. In: LaSalle, J. & Gauld, I.D., (Eds.), *Hymenoptera and Biodiversity*. C.A.B. International, Oxon, pp. 197-216.

Landis, D.A., 1994. Arthropod sampling in agricultural landscapes: ecological considerations. In: Pedigo, L.P. & Buntin, G.D., (Eds.), *Handbook of Sampling Methods for Arthropod Pests in Agriculture*. CRC Press, pp. 15-31.

Landis, D.A. & Haas, M.J., 1992. Influence of landscape structure on abundance and within-field distribution of *Ostrinia nubilalis* Hübner (Lepidoptera: Pyralidae) larval parasitoids in Michigan. *Environmental Entomology* **21**:409-416.

Landis, D.A. & Marino, P.C., 1999. Landscape structure and extra-field processes: impact on management of pests and beneficials. In: Ruberson, J., (Ed.), *Handbook of Pest Management*. Marcel Dekker Inc. NY, pp. 79-104.

Landis, D.A. & Menalled, F., 1998. Ecological considerations in the conservation of effective parasitoid communities in agricultural systems. In: Barbosa, P., (Ed.), *Conservation Biological Control*. Academic Press, CA, pp. 101-121.

Lawton, J.H., 1978. Host-plant influences on insect diversity: the effects of space and time. In: Mound, L.S. & Waloff, N., (Eds.), *Diversity of Insect Faunas*. Symposium of the Royal Entomological Society, London 9, pp. 105-125.

Lawton, J.H., 1983. Plant architecture and the diversity of phytophagous insects. *Annual Review of Entomology* **28**:23-39.

Lawton, J.H. & Schroder, D., 1977. Effects of plant type, size of geographical range and taxonomic isolation on number of insect species associated with British plants. *Nature* **265**:137-140.

Leius, K., 1963. Effects of pollens on fecundity and longevity of adult *Scambus buolianae* (Htg.) (Hymenoptera: Ichneumonidae). *Canadian Entomologist* **95**:202-207.

Leius, K., 1967. Food sources and preferences of adults of a parasite, *Scambus buolianae* (Htg.) (Hymenoptera: Ichneumonidae). *Canadian Entomologist* **99**:865-871.

Marino, P.C. & Landis, D.A., 1996. Effects of landscape structure on parasitoid diversity and parasitism in agroecosystems. *Ecological Applications* **6**:276-284.

Merriam, G., 1988. Landscape dynamics in farmland. *Trends in Ecology and Evolution* **3**: 16-20.

Murdoch, W.W., Evans, F.C. & Peterson, C.H., 1972. Diversity and pattern in plants and insects. *Ecology* **53**:819-829.

Murdoch, W.W., Chesson, J. & Chesson, P.L., 1985. Biological control in theory and practice. *American Naturalist* **125**: 344-366.

Powell, W., 1986. Enhancing parasitoid activity in crops. In: Waage, J. & D. Greathead, D. (Eds.),. *Insect Parasitoids*. Academic Press, Orlando, pp. 319-340.

Price, P.W., 1991. Evolutionary theory of host and parasitoid interactions. *Biological Control* **1**: 83-93.

Price, P.W., 1994. Evolution of parasitoid communities. In: Hawkins, B.A. & Sheehan, W., (Eds.), *Parasitoid Community Ecology*. Oxford University Press, pp. 473-491.

Rahim, A., Hashmi, A.A. & Khan, N.A., 1991. Effects of temperature and relative humidity on longevity and development of *Ooencyrtus papilionis* Ashmead (Hymenoptera: Eulophidae), a parasite of the sugarcane pest, *Pyrilla perpusilla* Walker (Homoptera Cicadellidae). *Environmental Entomology* **20**:774-775.

Riechert, S.E. & Bishop, L., 1990. Prey control by an assemblage of generalist predators: spiders in garden test systems. *Ecology* **71**:1441-1450.

Ryszkowski, L. & Karg, J., 1991. The effect of the structure of agricultural landscape on biomass of insects of the above-ground fauna. *Ekologia Polska* **39**:171-179.

Ryszkowski, L., Karg, J., Margalit, G., Paoletti, M.G. & Zlotin, R., 1993. Above ground insect biomass in agricultural landscapes of Europe. In: Bunce, R.B.H., Ryszkowski, L. & Paoletti, M.G (Eds.), *Landscape ecology and agroecosystems*. Lewis, Ann Arbor, Michigan, USA, pp. 71-82

Sheehan, W., 1991. Host range patterns of Hymenopteran parasitoids of exophytic Lepidopteran folivores. In: Bernays, E. (Ed.), *Insect-Plant Interactions Vol. III*. CRC Press, Boca Raton, pp. 209-247.

Sheehan, W., 1986. Response by specialist and generalist natural enemies to agroecosystem diversification: a selective review. *Environ. Entomol.* **15**: 456-461.

Sheehan, W., 1994. Parasitoid community structure: effects of host abundance, phylogeny, and ecology. In: Hawkins, B.A. & Sheehan, W. (Eds.) *Parasitoid Community Ecology*. Oxford University Press, pp. 90-107.

Southwood, T.R.E., Brown, V.K. & Reader, P.M., 1979. The relationship of plant and insect diversities in succession. Biological *Journal of the Linnean Society* **12**:327-348.

Stilling, P., 1990. Calculating the establishment rates of parasitoids in classical biological control programs. *American Entomologist* **36**: 225-230.

Stinson, C.S.A. & Brown, V.K., 1983. Seasonal changes in the architecture of natural plant communities and its relevance to insect herbivores. *Oecologia* **56**:67-69.

Szentkiralyi, F. & Kozar, F., 1991. How many species are there in apple insect communities?: testing the resource diversity and intermediate disturbance hypotheses. *Ecological Entomology* **16**:491-503.

Townes, H., 1962. Host selection patterns in Nearctic Ichneumonids. Proceedings of the 2nd International Congress of Entomology 2, 738-744.

Townes, H., 1972. Ichneumonidae as biological control agents In: *Proceedings of the Tall Timbers conference on ecological animal control by habitat management*. Tall Timbers Research Station, Tallahassee, Fl, pp. 235-248.

van Alphen, J.J.M. & Vet, L.E.M., 1986. An evolutionary approach to parasitoid host finding and selection. In: Waage, J. & D. Greathead, D. (Eds.), *Insect Parasitoids*. Academic Press, Orlando, pp. 23-54.

van Emden, H.F., 1963. Observations on the effects of flowers on the activity of parasitic Hymenoptera. *Entomologists Monthly Magazine* **98**:265-270.

van Emden, H.F., 1965. The role of uncultivated land in the biology of crop pests and beneficial insects. *Scientific Horticulture* **17**:121-136.

Wilcove, D.S., McLellan, C.H. & Dobson, A.P., 1986. Habitat fragmentation in the temperate zone. In: Soulé, M., (Ed.), *Conservation Biology: The Science of Scarcity and Diversity*. Sinauer Associates, Sunderland, MA, pp. 237-256.

CHAPTER 12

THE IMPACT OF FIELD BOUNDARY HABITATS ON THE DIVERSITY AND ABUNDANCE OF NATURAL ENEMIES IN CEREALS

P. DENNIS
Macaulay Land Use Research Institute, Craigiebuckler, Aberdeen Scotland, UK

G. L. A. FRY
Norwegian Institute for Nature and Cultural Research, Oslo, Norway

A. ANDERSEN
Norwegian Crop Research Institute, Ås, Norway

12.1 Introduction

Semi-natural biotopes of agricultural landscapes take the form of linear and insular structures which remain in predominantly cultivated areas (Lubbe, 1988). The linear structures are typically grass boundaries or hedges between fields, along farm tracks, roadsides, drainage ditches, water courses and forest edges (Greaves & Marshall, 1987). These semi-natural, remnant biotopes have been included in recent ecological studies of agricultural ecosystems because they provide habitat for farmland gamebird species (Potts, 1980; Sotherton, 1991), wildlife, in particular, song birds (Parish *et al.*, 1994) and butterflies (Dover, 1991). They were also considered to influence the species composition and population size of natural enemies in arable fields which were either the stenophagous predators or parasitoids of crop pests (van Emden, 1965) or polyphagous, generalist predators, which predate on crop pests only as part of a general diet which is typically composed of arthropods, mycoflora and herbage (Sotherton, 1985; Coombes & Sotherton, 1986). It was suggested that the densities of these predators of insect pests in crop fields may be increased by these adjacent habitats because they could provide shelter, breeding sites and sources of alternative food (Hagen *et al.*, 1976).

Beyond these adjacent boundary to field interactions, it was also proposed that the spatial arrangement of linear networks of these semi-natural biotopes in the agricultural landscape could also affect predatory arthropod diversity (Forman & Baudry, 1984; Mader, 1988). The traditional landscape and natural biodiversity resulting from earlier farming methods was characterised by smaller fields with correspondingly complex field boundary networks; semi-natural habitats in which populations of native plants and animals have been sustained in a predominantly agricultural landscape (Pollard *et al.*, 1974). An understanding of large scale processes is required to predict the consequences of

B. Ekbom, M. Irwin and Y. Robert (eds.), Interchanges of Insects, 195-214

the progressive loss of these biotopes due to modern agricultural practices in Europe (Fry, 1991).

This paper reviews the applied ecological research concerned with the value of field boundary habitats as refugia for natural enemies of arable pests but also discusses the contribution these predatory species make to general arthropod diversity on farmland (see also Dennis & Fry, 1992). Previously unpublished data are presented from a study of the effects of boundary habitats on the species composition and abundance of natural enemies of arable pests in the intensive arable farmland of Akershus county, southeast Norway. Spring-sown oats or barley are cultivated, usually with no rotation, after the stubble of the previous season's cereal crop is ploughed in the autumn. The study investigated the seasonal influence on natural enemy species and abundance of grass or shrub boundaries between fields and neighbouring fields, farm tracks or a forest edge, employing soil core sampling in winter and gutter trap and quadrat/visual search along boundary-field transects in summer (see Dennis, 1991; Dennis & Fry 1992; Dennis *et al.*, 1994).

12.2 Habitat Value of Field Boundaries for Predatory Arthropods

12.2.1 HABITAT REQUIREMENTS FOR WINTER REFUGIA

Grassy raised banks supported the highest overwintering densities of predatory arthropods out of four types of field boundaries studied in southern England (Sotherton, 1985). The planting of new grass strips within existing cereal fields also encourages high densities of overwintering predators (Thomas & Wratten, 1988; Thomas *et al.*, 1992). A higher proportion of populations of *Tachyporus hypnorum* (Staphylinidae) (Fig. 1) and *Demetrias atricapillus* (Carabidae) survived the winter when enclosed with vegetation of greater structural complexity on a field boundary of uniform structure (Dennis *et al.*, 1994). Sheltered, dry microhabitats in the upper 15 cm of soil were the main requirements for the survival of predators through winter and tussock grasses, in particular Cocksfoot, *Dactylis glomerata* created such favourable microhabitats (Luff, 1965; Thomas *et al.*, 1992; Dennis & Fry, 1992; Dennis *et al.*, 1994). Conversely, boundaries consisting of the pernicious agricultural weed, couch grass, where it was present in dense stands, supported similar high densities of natural enemies to those sampled in neighbouring semi-natural grass boundaries (Lageröf & Wallin, 1993).

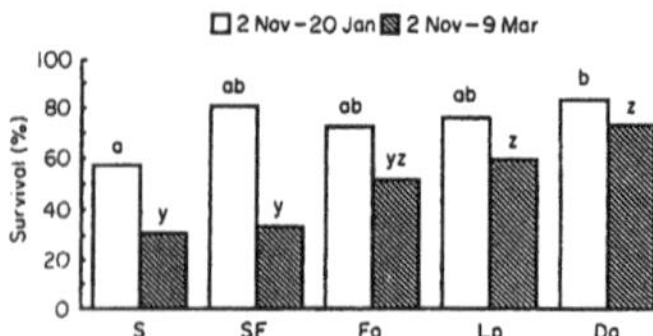

Figure 1. Mean percentage live *Tachyporus hypnorum* surviving two winter periods from November 1987 to January 1988 ($P < 0.05$, 23 % loss between S and Dg) and March 1988 ($P < 0.001$, 33-47 % loss between S, SF and Lp, Dg) on five experimental treatments of S, bare soil; SF, flint covered soil; Fo, *Festuca ovina*; Lp, *Lolium perenne*; and, Dg, *Dactylis glomerata* (after Dennis *et al.*, 1994).

The physical and biotic structure of the boundary habitat accounted for only 15-20% of the variation in the winter abundance of Coleoptera species within a general survey of grass boundaries and it was concluded that other factors, such as pre-winter crop husbandry, food supply and parasitism, may affect the dispersal power, habitat selection and cold hardiness of beneficial arthropod species within available boundaries and account for the observed variation in beetle numbers (Dennis *et al.*, 1994). Within - boundary habitat selection by predatory Coleoptera species has been identified and species, namely *Bembidion* spp. overwintered at the edge of the boundary, whereas other species, namely *Amara* spp. and *Tachyporus* spp. were found to overwinter throughout the boundaries. Higher densities of field-inhabitating predators were also found throughout the field margin ecotone during winter (Riedel, 1995).

12.2.2 SUMMER BOUNDARY HABITAT REQUIREMENTS OF NATURAL ENEMIES

The role of field boundaries is significant in winter for providing refugia but in summer they continue to influence the distribution of natural enemies in the cereal fields because they can generate a favourable zone of microclimate in the headlands which creates an ecotone effect (Bauer, 1989; Gourov, 1994) caused by short incursions of grassland arthropods into cereal fields from the boundaries (Dennis & Fry, 1992; Duelli & Obrist, 1995). Adult hoverflies use the nectar of wild flowers which are situated in some boundaries as a source of energy that allows them to search further into cereal fields for aphid populations. Sources of pollen in the boundary can improve the oviposition rates in the proximity of aphid populations (Cowgill *et al.*, 1993a, b). Sugars obtained from flower nectar can increase longevity, fecundity and searching activity for parasitoids which enhances the response of various parasitoid populations to rising pest populations (Gurr *et al.*, 1997). Recognized mechanisms for field boundaries in the crop season include a source of plants which serve as non-host food sources, non-crop vegetation as a habitat for alternative hosts / prey and the provision of shelter (Gurr *et al.*, 1997). Landis and Haas (1992) recognized that a field boundary provides shelter both as crop-season habitats and moderated microclimates. These crop-season interactions have been dealt with thoroughly in other parts of this volume (see Marino and Landis, Chapter 11).

12.2.3 EFFECTS OF CHANGES IN MANAGEMENT OF THE FIELD MARGIN

The ecological research concerned with farmland gamebirds highlighted the importance of the management applied to the area of crop adjacent to the boundary (the crop margin or headland). Removal of pesticide sprays increased plant cover and insect food abundance for Grey partridge chicks (Sotherton, 1991). Unsprayed headlands supported significantly higher densities of predatory arthropod species, including those recorded as feeding on aphid pests, and also contained higher densities of their alternative prey species (Chiverton & Sotherton, 1991). As a consequence of this modification of spray management, fewer cereal aphids were consumed in the headlands. Furthermore, cultivated but uncropped strips in the headland, which later developed a diverse cover of broadleaved plants, contained more species and a greater abundance of carabids than cropped headlands that were sprayed or unsprayed (Hawthorne & Hassall, 1995). This raised the question whether predators would remain in this more favourable habitat and not disperse into the adjacent cereals to impose a predation pressure on cereal pests. However, further research to address this question showed that at the distance of 8 m into the crop significantly higher densities of these predators and correspondingly lower densitites of cereal aphids were found adjacent to the uncropped headlands than adjacent to either sprayed or unsprayed headlands (Hawthorne & Hassall, 1995).

12.3 Distribution of Predatory Arthropods in Summer

The interactions of natural enemies between semi-natural habitat and cultivated areas are characterized by six types of distribution (Table 1, after Duelli & Obrist, 1995). Examples of species for each of these distribution types can be found in Dennis (1991), Duelli & Obrist (1995), Hawthorne & Hassall (1995), Kromp & Nitzlader (1995) and Riedel (1995). It was estimated that 60% of the species of distribution types one to three were dependent on semi-natural habitats. However, the role of boundary habitats as winter refugia was not taken into account and the stated proportion may be an under-estimate when these seasonal effects are taken into account (Duelli & Obrist, 1995). Dennis and Fry (1992) further divided predatory arthropods of distribution types two to four into seasonal categories, resident through winter into summer (four spp.), present from the establishment stage of aphid population growth (ten spp.) and present from the exponential stage of aphid population growth (sixteen spp.).

Table 1. Six distribution types related to the interaction of predatory arthropods with semi-natural habitats and cultivated areas, indicating the total number of predatory species of each type (after Duelli & Obrist, 1995)

Distribution	Description	Number of species
Type 1	Distribution only in semi-natural habitats	34
Type 2	Distribution maximum in semi-natural habitat, diminishing into the cultivated area	136
Type 3	A narrow distribution with a maximum close to the field boundary	60
Type 4	Ubiquitous species	43
Type 5	Distributions exclusively in cultivated areas	38
Type 6	Species with no affinities to habitats across agricultural landscapes, randomly distributed	8

12.3.1 SPRING DISPERSAL PATTERNS

The interactions of predatory arthropods between semi-natural habitats and arable land are reviewed and further data are analysed to determine the extent of the influence of field boundary habitats on natural enemy populations within adjacent fields, through the spring and summer growing season. The main objective of the experimental work was to investigate the roles of field boundary habitats in supporting a larger natural enemy assemblage in adjacent cereal fields. This was achieved by measuring the:

* species composition in boundary habitats during winter, assessing the similarity with that of the early spring field assemblage of predators in spring, and
* spring/ summer field distribution patterns of predators where distribution patterns can infer dispersal processes.

Emphasis was placed on the taxa which contributed most polyphagous predatory species to the boundary and cereal field habitats, the ground beetles (Coleoptera: Carabidae) and rove beetles (Coleoptera: Staphylinidae) but data on the stenophagous predators, ladybirds (Coccinellidae) and syrphid larvae (Diptera: Syrphidae) and other polyphagous predators, for example, soldier beetles (Coleoptera: Cantharidae) and money and wolf spiders (Araneae: Linyphiidae and Lycosidae) were also recorded.

(a) Distribution of predatory arthropods along boundary-field transects

The spring/ summer distribution of arthropods was sampled along the boundary-field transects with quadrats and visual search before sowing and directional gutter traps after cereals were sown. The directional gutter traps are described in Dennis (1991) and Dennis and Fry (1992) and were composed of a section of plastic guttering, dug into the ground so that only one side was flush with the soil. The other side was raised above the ground such that ground walking arthropods could not be captured from that side. Fluon was painted on to the plastic inside the guttering to prevent escapes. The guttering was tilted slightly to direct captured individuals towards a down pipe and plastic cup filled with preservative. A total of 36 traps were placed at six distances along six transects out into cereal fields at two different boundaries.

A significant difference was found in the distribution of predatory species along boundary/ field transects from initial spring activity as sampled by quadrat and visual counts until sowing, and from May until the cereal crop ripened towards the end of July, as sampled by gutter traps (Fig. 2). Individual predators showed a similar but non-significant pattern of distribution as the season progressed (Fig. 3). The data on individual predator abundance collected using directional gutter traps were analysed with two-way ANOVA for distance along transect and direction of catch. There were sixteen species of carabids, staphylinids and spiders selected by their abundance in the gutter trap transects of which seven showed significant trends in the distance or direction of capture (Table 2). *Carabus nemoralis, Bembidion guttula, Stenus biguttatus, B. lampros, B. quadrimaculatum, Trechus secalis* and *Pardosa* sp. had significantly higher captures in traps positioned closer to the field boundary (Table 2). However, *Carabus nemoralis* was restricted to traps at 5 m from the boundary and provides an example of a grassland species which wanders into arable headlands from the grassy boundary habitat, hence the field margin represents an ecotone beyond which this species is not active.

Patterns of spatial dependence of the abundant species were calculated using Moran's i statistic of spatial dependence for each distance class in the 36 trap grid of gutter traps up to 50 m from the field boundary (see Liebhold *et al.*, 1993). Moran's i indicates positive (+) or negative (-) spatial autocorrelation at specific distance classes and can be recalculated for specific directions. Positive values indicate that there is consistent similarity of catch between traps at a specific distance class. Negative values likewise indicate that there is a trend of inverse trap abundance between traps at a specific distance class. These analyses were used to indicate the presence of a spatial structure in the species data, independent of the actual size of catch. An analysis was undertaken of the spatial dependence of beetles listed in Table 2 compared with half normal or exponential decay distributions from boundaries into fields. For a beetle species to fit these distribution models, spatial dependence must be significant and positive at distance classes of 4, 8 and 12 m (i > +0.228; P < 0.05) and significant and negative at distance classes of 20 and 24 m (i < -0.341; P < 0.05).

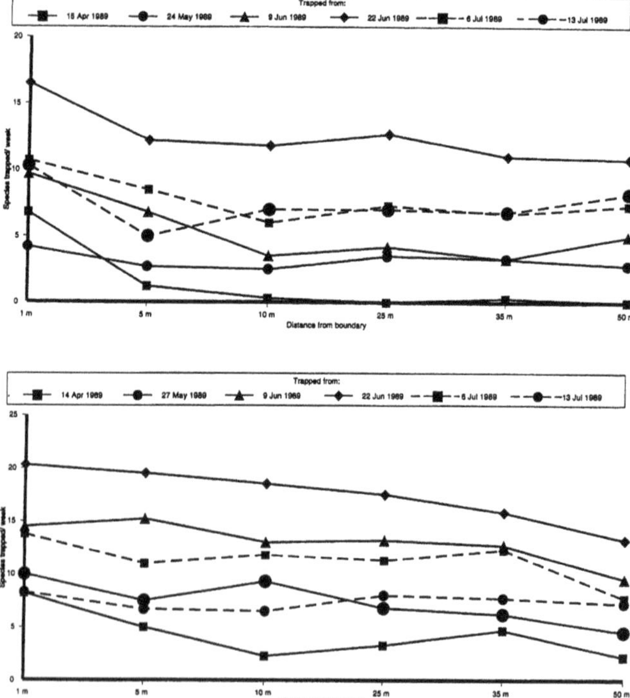

Figure 2. Distribution of natural enemy species along boundary-field transects (a. Ski and b. Huseby sites). Differences tested with two-way analysis of variance for date and distance: a. date: $F_{5,215} = 41.45$, $P < 0.001$; distance: $F_{5,215} = 18.59$, $P < 0.001$. b. date: $F_{7,287} = 67.84$, $P < 0.001$; distance: $F_{5,287} = 11.13$, $P < 0.001$.

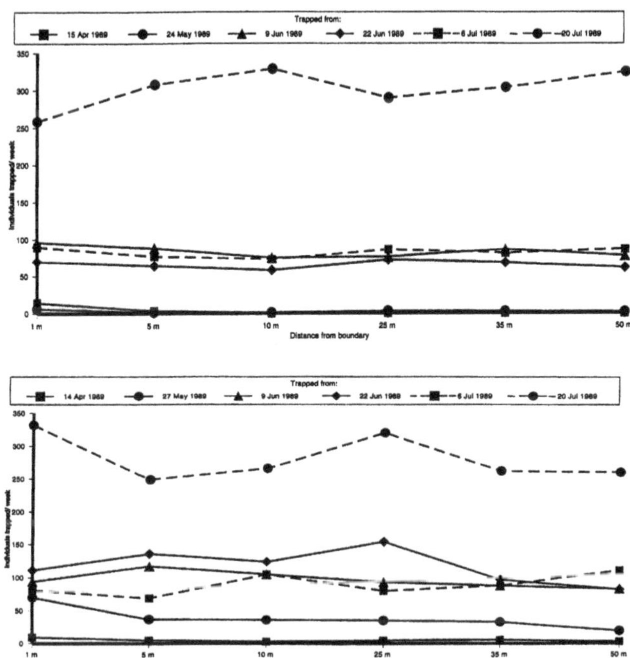

Figure 3. Distribution of individual predators along boundary-field transects (a. Ski and b. Huseby sites). Differences tested with two-way analysis of variance for date and distance: a. date: $F_{3,143} = 182.26$, $P < 0.001$; distance: $F_{5,143} = 0.08$, not significant. b. date: $F_{4,150} = 133.87$, $P < 0.001$; distance: $F_{5,150} = 1.65$, not significant.

T. secalis, *B. lampros*, *B. guttula* and *B. quadrimaculatum* showed patterns of spatial dependence consistent with the hypothetical distributions out from field boundaries (Table 2). *A. rugosus*, *S. biguttatus* and *L. fulvipenne* showed no significant spatial dependence and the patterns of catch were therefore random. *C. nemoralis* and *Pardosa* sp. showed positive spatial dependence at the shortest distance classes only because the activity of these species was restricted to the field margin whereas several species, *e.g. P. melanarius* and *A. plebeja*, showed spatial dependence at distance classes not consistent with the hypothetical linear distributions but indicating spatial aggregations within the field at different spatial scales (Table 2). There may also be influences on the activity of these species from the location of other field boundaries, or partial dispersal from the field boundary into the field margin, giving spatial association at short distances, associated with later flight dispersal which would cause random patterns of distribution at larger spatial scales. This type of behaviour was recorded for *Tachyporus* spp. and *Philonthus* spp. during behavioural studies in laboratory arenas (Dennis & Sotherton, 1994).

Table 2. Predatory arthropods which occurred in abundance in directional gutter traps, indicating overwintering location, spring distribution type, species marked * all showed significantly higher abundances in the field margin (Dennis & Fry, 1992)

Species	Order: Family	Overwintering location	Distribution type - winter	Distribution type - summer
Carabus nemoralis 3	Coleoptera: Carabidae	Field boundary, grassland soil	1	1
Pterostichus melanarius 3	Coleoptera: Carabidae	Boundary and cultivated soil	4	4
Clivina fossor 3	Coleoptera: Carabidae	Boundary and cultivated soil	4	4
*Trechus secalis*1*	Coleoptera: Carabidae	Field boundary soil	1	2
*Bembidion guttula*1*	Coleoptera: Carabidae	Field boundary vegetation	1	2
*Bembidion lampros*1*	Coleoptera: Carabidae	Field boundary vegetation	1	2
*Bembidion 4-maculatum*1*	Coleoptera: Carabidae	Field boundary vegetation	1	2
Amara plebeja 3	Coleoptera: Carabidae	Field boundary soil	1	4
*Stenus biguttatus*2*	Coleoptera: Staphylinidae	Field boundary vegetation	1	4
Lathrobium fulvipenne 2	Coleoptera: Staphylinidae	Boundary and cultivated soil	4	4
Anotylus rugosus 2	Coleoptera: Staphylinidae	Field boundary vegetation	1	4
Tachyporus hypnorum 4	Coleoptera: Staphylinidae	Field boundary vegetation	1	5
Tachyporus dispar 4	Coleoptera: Staphylinidae	Field boundary vegetation	1	4
Tachyporus obtusus 4	Coleoptera: Staphylinidae	Field boundary vegetation	1	4
Erigone atra 2	Araneae: Linyphiidae	Boundary and cultivated soil	4	5
*Pardosa sp.*3*	Araneae: Lycosidae	Field boundary vegetation	1	2

Species (1) showed patterns of spatial dependence consistent with a half normal or exponential decay distribution out to 50 m from field boundaries (Moran's i > +0.12, P < 0.05 at shorter distance classes; i < -0.17, P < 0.05 at larger distance classes. Species (2) showed no significant spatial dependence at any distance class (P < 0.05). Species (3) showed spatial dependence at particular distance classes but not consistent with the hypothetical, linear distributions. There were too many zeros in the data matrices of the *Tachyporus* spp. (4) to calculate Moran's i statistic.

(b) Relationship with crop growth stage and soil temperature

Infra-red emissions (temperature) and crop development were recorded along the boundary-field transects for comparison with the data collected by quadrat/visual search and gutter traps. There was a significant difference in the radiated temperatures of the soil and vegetation along boundary/field transects from early spring (Fig. 4). The boundary temperatures were significantly warmer in April and May, and significantly cooler in June. This suggested a buffering effect on temperature fluctuations. The late morning recordings showed during cold weather, that the boundaries maintained a higher temperature, and that whilst the season progressed, the boundaries were cooler, hence they were slower to heat up. During this time, the field remained ploughed until 20 April when it was harrowed until 20 May, when tilling of cereals took place (Table 3). The crop cover developed from

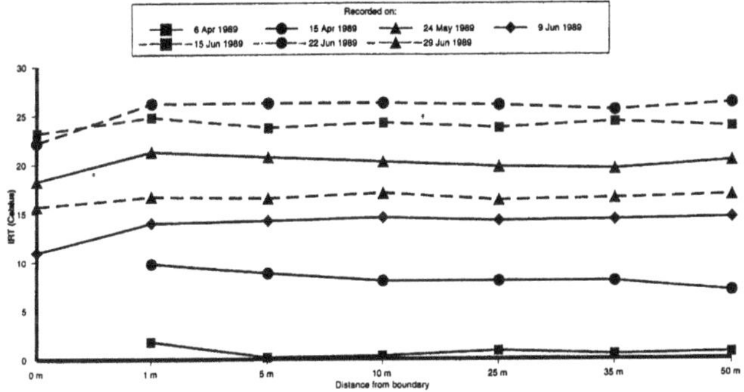

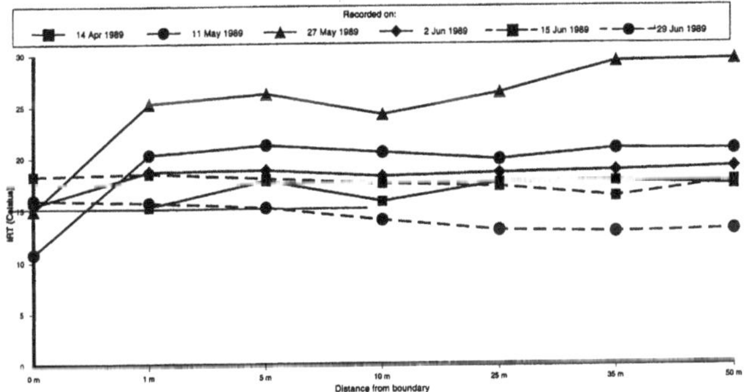

Figure 4. Infra red temperatures along boundary-field transects (a. Ski and b. Huseby sites) Analysis of variance of distance classes for each date ($n_1 = 5$, $n_2 = 35$): a. Ski: $F_{6\ April} = 3.11$**; $F_{15\ April} = 17.24$**; $F_{24\ May} = 0.11$ns; $F_{9\ June} = 14.97$***; $F_{15\ June} = 2.43$*; $F_{22\ June} = 3.66$**; $F_{29\ June} = 0.35$ns; b. Huseby: $F_{14\ April} = 1.02$ns; $F_{11\ May} = 7.94$***; $F_{27\ May} = 38.71$***; $F_{2\ June} = 3.18$*; $F_{15\ June} = 1.66$ns; $F_{29\ June} = 8.27$***.

Feekes scale 1 (7.6% cover) on 25 May and reached Feekes scale 8 (85% cover) by 29 June, the development stage where maximum crop cover was achieved (Table 3).

The number of species of predators active in the transects of gutter traps were significantly related to the stage of crop development and radiated temperature measurements (Table 3). However, the overall number of predatory individuals related significantly only to crop growth stage (Table 3). These statistics could suggest there was a favourable microclimate zone for more species closer to the field boundary but this could also be interpreted as an edge effect where grassland species resident in the boundary spill over into field margin. The stronger relationship of individuals and species with crop development suggests a requirement for vegetation cover before dispersal into the field but a temporal correlation could also account for this pattern. For instance, the progressive increase in activity and dispersal into fields after the physiological development (winter diapause) of the predatory species may relate to the rise in ambient temperature that also determines crop germination, growth and crop cover.

Table 3. Crop development and mean irradiated temperature on the six field boundary - field transects *e.g.* the Ski site. Number of predatory arthropod species significantly related to the crop stage of development and radiated temperature: $R = 0.54$, $a = 3.55$, $b_{Feekes} = 0.93$, $b_{temp} = 0.13$, $F_{2,358} = 7.47$, $P < 0.001$, $R^2 = 0.30$. Total predatory individuals related significantly only to crop growth stage: $R = 0.46$, $a = 12.8$, $b_{Feekes} = 4.23$, $F_{2,358} = 4.66$, $P < 0.001$, $R^2 = 0.20$

Date	Feekes index	Radiated temperature (°C)	Stem density (m^{-2})	% cover
20/4-89	0 (harrowed)	-0.2	0	0
25/5-89	1 (germination)	28.3	185	7.6
9/6-89	2	13.9	154	27.7
15/6-89	5	23.4	152	40
22/6-89	7	25.4	167	65
29/6-89	8	16.2	157	85
15/8-89	10 (harvest)	-	160 - stubble	85

(c) Predation pressure along boundary-field transects

On the transects, there was a weak relationship between total individual predators and baits removed (Dennis, 1991) but this reflects the differences in intake rate of the individuals of different predatory species and their specialization for aphids as prey. For instance, *Tachyporus* spp. consume few aphids as a proportion of their overall insect and fungal diet in cereal fields (Dennis *et al.*, 1991) whereas *Adalia septempunctata* larvae have a high intake of solely aphids. In an experiment which altered the population sizes of natural enemies in oats by the application of different rates of the insecticide, dimethoate, predation rates of aphid baits related significantly to the differences in overall densities of natural enemies within each treatment (Dennis *et al.*, 1993). Aphid populations were also shown to be lower in crops adjacent to cultivated headlands where natural enemy densities were higher (Hawthorne & Hassell, 1995), a similar correlation has previously been identified in conventional cereal fields and low aphid and higher natural enemy densities were associated with headlands compared with open field locations (Chambers *et al.*, 1982).

12.3.2 WINTER BOUNDARY AND SUMMER FIELD SPECIES COMPOSITION OF PREDATORS

Overwintering arthropods were extracted from soil cores removed from eight field boundaries using Tullgren funnels (Dennis *et al.*, 1994). The spring arthropod distribution was sampled along boundary-field transects with visual search within placed quadrats prior to the sowing date of cereals. It was found that approximately 60% of the overwintering predatory species were active in the field from the analysis of the species sampled from the boundaries in winter and cereal fields in spring and summer (Table 4). The species from the boundary contributed only 39% of the species assemblage sampled in the field. However, the distribution of species across the field was skewed and 28% more species occurred within 10 m of field boundaries (Dennis & Fry, 1992). The field assemblage of beetles included species which overwintered in the ploughed stubbles, for example, *Lathrobium fulvipenne* (Staphylinidae), *Clivina fossor* (Carabidae) and *Erigone atra* (Linyphiidae).

Table 4. Natural enemy species associations of field boundaries in winter and cereal fields in spring. Association Analysis was carried out on the species presence/ absence in the field or boundary at seven locations. There was an overall significant and positive association ($V = 3.57$; $W = 49.95$; $P < 0.05$) although only four of the seven locations were grouped by species composition in the boundaries and adjacent fields

Location	Ski			Aas	Huseby		Kroer
Boundary type	track-field grass	field-field shrub	field-field grass	road-field grass	field-field grass	grass around rock outcrop	forest-field grass
Boundary - winter	15	19	21	28	18	19	21
Field - spring	25	13	55	34	68	12	37
Species shared	8	5	17	17	17	7	14

A large component of the species assemblage, particularly staphylinid species that disperse rapidly by flight, immigrate from more remote semi-natural habitats either in the general network of boundaries or from fields of grass and permanent pastures. About 29 out of 219 species, for example *Tachyporus* spp. and *Philonthus* spp., move into cereal fields in summer for reproduction (Duelli & Obrist, 1995). A further ten species were more nomadic, moving by regular flights between field habitats in summer, not just from winter hibernation sites in semi-natural vegetation to the field habitats for breeding in a distinct migration in spring and back in the autumn, *e.g. Arpedium quadrum* and *Platystethus arenarius* (Good & Giller, 1991; Duelli & Obrist, 1995).

Overwintering larvae create a source of error because they were not identified in this study and thus may have caused an underestimate of the contribution of the arthropod assemblage in the boundary refugia to the spring field assemblage, *e.g. Trechus secalis* and *Pterostichus melanarius*. Further, the glacial, insular grass habitat and shrub boundary sites were affected by nests of wood ants, where the species and abundance of natural enemies and probably aphid and other pests also were diminished compared with other boundary sites (Table 4).

In summary, boundary habitats have a seasonal and spatial influence on the natural enemy populations present in cereal fields during the spring and summer. Approximately 60% of the overwintering species in field boundaries are active in cereal fields in spring and summer. The species from boundary habitats contribute 39% of the predatory species active in the cereal fields during the growing season, although the distribution is skewed and 28% more species occupy the outer 10 m of cereal fields (Dennis & Fry, 1992).

12.4 Landscape Pattern of Field Boundaries

12.4.1 DISTRIBUTION OF PREDATORS ALONG BOUNDARY NETWORKS

The summer distribution of carabid beetles along boundaries and their inter-connections (nodes) was also sampled with pitfall traps. The contribution of these components of the boundary network to the species and trap abundance of ground beetles were analysed using multivariate and regression statistics. Pitfall samples of carabid beetles were taken at junctions (nodes) and mid-points of field boundaries.

A comparison by Wilcoxon signed rank test of the number of species found exclusively in nodes and species found exclusively in boundaries showed that the species assemblages were similar across the 21 field boundaries investigated ($Z = -1.40$, ns). A regression analysis of node against boundary species, where the intercept was forced to the origin was significant with the slope close to one ($b = 1.097$, $F_{1,22} = 10.15$, $P < 0.01$). This was evidence of an equal distribution of species along networks of predominantly grass boundaries. There was, however, considerable variation in the number of species sampled at each of the 22 paired sites, range 3-19. A factor analysis reduced the species-abundance data from the combined pitfall traps of each location to three factors which accounted for 96% of the variation in the data. Five species correlated significantly with these factors but there was found to be no significant difference in abundance of *Pterostichus niger*, *A. plebeja* and *B. quadrimaculatum* between boundary and nodes. However, *T. secalis* and *L. pilicornis* were captured in significantly higher abundance in nodes compared with boundaries ($t = 2.29$, $P < 0.05$ and $t = 2.87$, $P < 0.05$, respectively). *T. secalis* correlated with ordination factor one ($r = 0.95$) and *L. pilicornis* correlated with ordination factor two ($r = 0.74$).

12.4.2 SPATIAL PROCESSES AFFECTING PREDATORY ARTHROPODS IN BOUNDARY NETWORKS

(a) Boundaries and the movement of natural enemy populations between fields

Ecotoxicological research has modified models of metapopulation dynamics to predict the effects on the population of a natural enemy of asynchronous applications of pesticides onto different crop fields (Sherratt & Jepson, 1993). For the metapopulation model to be applicable in this situation, the field boundary must act as a semi-permeable barrier to the movement of individuals of the natural enemy sub-population, hence individuals in untreated fields will not be correlated with the same mortality effect (Boois & Nijs, 1992). The delay in the re-invasion from unsprayed fields to a field in which the crop has recently been sprayed would create a more stable natural enemy (meta-)population at the farm-scale (Sherratt and Jepson, 1993). Boundaries of high permeability to the movement of individuals of the natural enemy would not provide a sufficient isolation of each fields' sub-population from the pesticide treatment of an individual field so that there would be a correlated mortality in all sub-populations. Boundaries of low permeability would not contribute to the recovery of the sub-population of an individual treated field, and likewise, would not receive individuals from neighbouring fields after pesticide application. The overall population would be reduced by sequential applications of pesticides to each of the fields under these conditions. Similarly, for a boundary of fixed permeability, species with high or low dispersal powers would be most likely to have unstable population dynamics.

The movement of larger carabid species can be moderately retarded by grassy boundaries (Frampton *et al.*, 1995) and hedgerows (Mauremootoo & Wratten, 1994) between cereal fields and gives credence to the application of metapopulation dynamics models to the cereal ecosystem. Five days after the release of marked individuals of the carabid, *Harpalus rufipes*, a 1.2 m wide, grass boundary was shown to reduce the permeability of the population from the c. 63% recorded in the barley crop to c. 44% (Fig. 5).

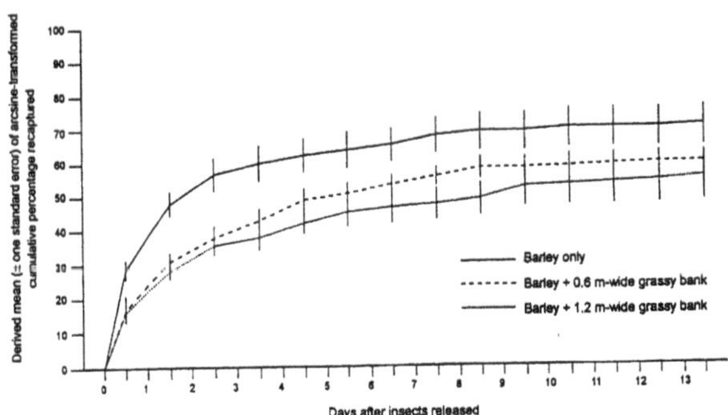

Figure 5. Effect of different widths of grassy bank on the rate of recapture of *Harpalus rufipes* in a barley crop (after Frampton *et al.*, 1995).

(b) Boundary networks and population dynamics of predatory arthropods

Intensification of arable crop production has caused the continued loss of elements of natural habitat in the agricultural landscape over recent decades (Fry, 1991). The reduction in the area and connectivity of remnant biotopes in agricultural landscapes has consequences for the population viability of plant and animal populations of nature conservation importance (Mader, 1988). Species may also be lost from areas designated as nature reserves if they depend on the connections afforded by the surrounding linear habitats (Saunders & Hobbs, 1989). In most European countries, the majority of wildlife species live in the farmed landscape, not in nature reserves, and it is imperative that populations of these species continue to coexist with changes in agricultural production (Anon., 1995). A variety of semi-natural biotopes contribute significantly as habitat to farmland biodiversity (Kaule & Krebs, 1989; Fry, 1991) and buffer the effects of cultivation and pesticide practices applied to crops (Dover *et al.*, 1990).

The spatial arrangement of boundary networks affects general arthropod diversity of the agricultural landscape (Burel, 1992; Petit & Burel, 1997). Boundaries have a role in the spread of woodland and grassland species into the wider farm landscape (Mader, 1988; Baudry, 1988). For increasing distances from farm woodlands, the species composition of ground beetles has a higher proportion of woodland species in boundaries composed of woody shrubs than simple grassy boundaries (Baudry, 1988). Conversely, networks of hedgerows inter-connect small woodlands in the farm landscape and contribute to the persistence of populations of woodland ground beetles (Charrier *et al.*, 1997; Petit & Burel, 1997). However, there is concern that the current distribution of some woodland ground beetle species may reflect the historic landscape structure before clearance of the hedgerow network related to agricultural intensification such that populations may no longer be viable (Petit & Burel, 1997). Computer models are being developed of the spatial dynamics of species, for instance, the ground beetles, to estimate the impact of changes in the area of remnant natural biotopes and extent of networks of field boundaries connecting them together in different agricultural landscapes (Rushton *et al.*, 1996). By recalculating such species-centred models for a number of different wildlife species, it should be possible to calculate critical values of natural habitat requirements to maintain general biodiversity on farmland.

12.5 Conclusions

Research has demonstrated a summer and winter influence of boundary habitats on the populations of a significant number of natural enemies of crop pests in arable fields (reviewed by Dennis & Fry, 1992). In winter, field boundary habitats provide winter refugia for polyphagous predators which are shown to feed on pest species in adjacent crops during summer. However, in summer they continue to influence the distribution of natural enemies across agricultural landscapes, providing either favourable zones of microclimate and habitat for species which will not disperse far beyond the field margin, nectar and pollen sources for hoverflies and parasitic wasps, or semi-permeable barriers to the movement between fields of natural enemy populations which stabilize the dynamics of populations in response to pesticide and cultivation practices at the farm scale. The same networks of boundaries support herbivorous and predatory insect species, *e.g.* butterflies and woodland carabids, which do not interact with the field crops but which contribute to the general arthropod biodiversity of agricultural landscapes.

Numerous species from several predatory taxa, namely, Carabidae, Staphylinidae, Coccinellidae, Cantharidae (Coleoptera), Syrphidae (Diptera), Chrysopidae (Neuroptera), Linyphiidae and Lycosidae (Araneae) are aphid predators in cereal crops in spring and summer (Sunderland *et al.*, 1987; Chiverton, 1987) and are known to overwinter as adults or larvae in grassy field boundaries (Van Emden, 1965; Sotherton, 1985; Thomas *et al.*, 1992; Wratten & Thomas, 1990; Coombes & Sotherton, 1986; Dennis *et al.*, 1994; Andersen, 1997). When *Trechus secalis*, *Pterostichus melanarius* and *Carabus nemoralis* were given a choice of woodland or cereal field habitats, the net movement pattern was towards the woodland, although the net movement of the adult overwintering species, *Bembidion lampros*, *Pterostichus cupreus*, *Agonum dorsale* and the larval overwintering *Harpalus rufipes*, was towards the cereal field (Wallin, 1986). The contribution of semi-natural biotopes to field populations of carabids (ground beetles) may have been overestimated (Riedal, 1991) because only a limited number of species, *e.g. Bembidion* spp. and *Agonum* spp., migrate into fields after using adjacent habitats as winter refugia (Table 4; Coombes & Sotherton, 1986; Kromp & Nitzlader, 1995). This suggests that investigations into the role of field boundaries must be balanced by developing a better understanding of the effect of autumn and winter field cultivation practices and chemical inputs on those carabid species which remain in the field throughout the year, either as adults or subterranean larvae (Kromp & Nitzlader, 1995).

In spring and summer, the species composition and abundance of natural enemies declines away from boundary habitats composed of semi-natural vegetation (Dennis, 1991; Dennis & Fry, 1992). Thus, the reduced network of field boundaries in agricultural regions with larger field systems may limit the distribution of beneficial arthropods into crops and this may have consequences for the growth rate of insect pest populations. Linear distribution patterns of predatory species were detected adjacent to boundaries for a number of predatory species (Fig. 2), but these distributions can theoretically be caused by different processes which may not be directly related to individuals overwintering in boundaries adjacent to crop fields (Table 5).

Table 5. Ecological determinants of an effect of field boundaries on the linear distributions of natural enemy species and individuals into field crops during spring and early summer

No.	Ecological determinant
1.	The dispersal pattern of predators progressively moving out from the winter refugia of the field boundary into the adjacent field
2.	The existence of a field-margin ecotone of grassland predators, active primarily in the boundary grass habitat
3.	The accumulation of immigrants within a favourable area for microclimate or prey in the field margin after they have dispersed randomly into the field from remote refugia

Many species of carabids do not move from the woody field boundaries into open fields but they do represent part of the general biodiversity on farmland (Baudry, 1988; Petit & Burel, 1997). Many species, like these woodland carabids, are restricted to the networks of semi-natural biotopes which remain within agricultural landscapes. A continued reduction in the network of boundaries or remnant biotopes that are represented on farmland will further reduce the biodiversity value of the agricultural landscape. This concern about the plight of biodiversity on farmland and the obligations of governments to honour the Convention on Biological Diversity (Anon., 1995) has altered the emphasis of research to consider the ecology of arthropod species which have a direct nature conservation value, *e.g.* butterflies (Dover, 1991) or species which provide food for other farmland wildlife of conservation interest, *e.g.* songbirds (Parish *et al.*, 1994) and gamebirds, *e.g.* the Grey partridge (Potts, 1980; Sotherton, 1991). There could be risks to general arthropod diversity in promoting natural enemies by the management or provision of boundary habitats designed to benefit particular beneficial species (Dennis & Fry, 1992; Lagerof & Wallin, 1993). A positive relationship was established for the density of overwintering natural enemies and general arthropod diversity in semi-natural grassland boundaries but this trend reversed at high predator densities (Fig. 6; Dennis & Fry, 1992). Boundaries consisting of dense stands of the pernicious agricultural weed, couch grass, supported high densities of natural enemies but couch grass was marked for its low general arthropod diversity compared with neighbouring semi-natural grass boundaries (Lagerof & Wallin, 1993). In conclusion, a balance must be struck between the aims of promoting populations of natural enemies in cereal fields by maintaining or augmenting grass boundaries and the wider requirements to conserve general biodiversity by protecting existing natural biotopes in the agricultural landscape.

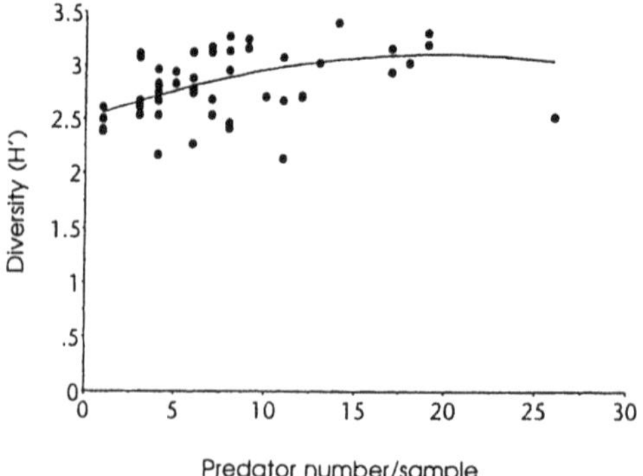

Figure 6. Total aphid feeding predators and general arthropod diversity (Shannon index H') in 52 soil samples (0.04 m^2) taken from six field boundaries in southeast Norway during winter 1990 ($r = 0.558$; $P < 0.001$; $a = 2.455 + 0.73b - 0.002b^2$; $F = 11.08$; $P < 0.01$) (after Dennis & Fry, 1992).

Acknowledgements

This work was funded by a Norwegian Agricultural Research Council (NLVF) fellowship at the Norwegian Institute of Nature Research and State Plant Protection Institute (1989-1990) situated at the Agricultural University of Norway, Ås. The pitfall trap surveys of boundaries and nodes was carried out by Chris Greaves (Biology Department, University of Southampton, UK). Comprehensive identification and sorting of the arthropods was carried out by Peter Dennis, Arild Andersen and John Mauremootoo (Biology Department, University of Southampton, UK). This chapter was prepared with financial assistance towards travel and subsistence from NINA-NIKU in 1997 which allowed further data analyses and collaboration between co-authors.

12.6 References

Andersen, A., 1997. Densities of overwintering carabids and staphylinids (Col., Carabidae and Staphylinidae) in cereal and grass fields and their boundaries. *Journal of Applied Entomology* **121**:77-80.

Anon., 1995. *Pan-European landscape and biodiversity strategy*, Council of the European Union, Strasbourg.

Baudry, J., 1988. Hedgerows and hedgerow networks as wildlife habitat in agricultural landscapes. In: Park, J.R., (Ed.), *Environmental Management in Agriculture, European Perspectives*, Belhaven Press, London, pp. 111-124.

Bauer, L.J., 1989. Moorland beetle communities on limestone habitat islands. I. Isolation, invasion and local species diversity in carabids and staphylinids. *Journal of Animal Ecology* **58**:1077-1098.

Booij, C.J.H. & den Nijs, L.J.F.M., 1992. Agroecological infrastructure and dynamics of carabid beetles. *Proceedings of Experimental and Applied Entomology* **3**:72-78.

Burel, F., 1992. Effect of landscape structure and dynamics on species diversity in hedgerow networks. *Landscape Ecology* **6**:161-174.

Chambers, R.J., Sunderland, K.D., Stacey, P.L. & Wyatt, I.L., 1982. A survey of cereal aphids and their natural enemies in winter wheat in 1980. *Annals of applied biology* **101**:175-178.

Charrier, S., Petit, S. & Burel, F., 1997. Movements of *Abax ater* Vill. (Coleoptera: Carabidae) in woody habitats of a hedgerow network landscape: a radiotracing study. *Agriculture, Ecosystems and Environment* **45**:133-144

Chiverton, P., 1987. Predation of *Rhopalosiphum padi* by polyphagous predatory arthropods during the aphids' pre-peak period in spring barley. *Annals of applied biology* **111**:257-269.

Chiverton, P. & Sotherton, N.W., 1991. The effects on beneficial arthropods of the exclusion of herbicides from cereal crop edges. *Journal of Applied Ecology* **28**:1027-1039.

Coombes, D.S. & Sotherton, N.W., 1986. The dispersal and distribution of polyphagous predatory Coleoptera in cereals. *Annals of applied biology* **108**:461-474.

Cowgill, S.E., Wratten, S.D. & Sotherton, N.W., 1993a. The effects of weeds on the number of hoverfly (Diptera: Syrphidae) adults and the distribution and composition of their eggs in winter wheat. *Annals of applied biology* **123**:499-514.

Cowgill, S.E., Wratten, S.D. & Sotherton, N.W., 1993b. The selective use of floral resources by the hoverfly *Episyrphus balteatus* (Diptera: Syrphidae) on farmland. *Annals of applied biology* **122**:223-231.

Dennis, P., 1991. The temporal and spatial distribution of arthropod predators of the aphids *Rhopalosiphum padi* (L.) and *Sitobion avenae* (F.) in cereals next to field margin habitats. *Norwegian Journal of*

Agricultural Sciences **5**:79-88.

Dennis, P., Wratten, S.D. & Sotherton, N.W., 1991. Mycophagy as a factor limiting predation of aphids (Hemiptera: Aphididae) by staphylinid beetles (Coleoptera: Staphylinidae) in cereals. *Bulletin of Entomological Research* **81**:25-31.

Dennis, P. & Fry, G.L.A., 1992. Field margins: can they enhance natural enemy population densities and general arthropod diversity on farmland? *Agriculture, Ecosystems and Environment* **40**:95-115.

Dennis, P., Fry, G.L.A. & Thomas, M.B., 1993. The effects of reduced doses of insecticide on aphids and their natural enemies in oats. *Norwegian Journal of Agricultural Sciences* **7**:311-325.

Dennis, P. & Sotherton, N.W., 1994. Behavioural aspects of staphylinid beetles that limit their aphid feeding potential in cereal crops. *Pedobiologia* **38**:222-237.

Dennis, P., Thomas, M.B. & Sotherton, N.W., 1994. Structural features of field boundaries which influence the overwintering densities of beneficial arthropod predators. *Journal of Applied Ecology* **31**:361-370.

Dover, J.W., Sotherton, N.W. & Gobbett, K., 1990. Reduced pesticide inputs on cereal field margins: the effects on butterfly abundance. *Ecological Entomology* **15**:17-24.

Dover, J.W., 1991. The conservation of insects on arable farmland. In: Collins, N.M. & Thomas, J.A. (Eds.), *The conservation of insects and their habitats*, Academic Press, London, pp. 293-318.

Duelli, P. & Obrist, M., 1995. Comparing surface activity and flight of predatory arthropods in a 5 km transect. In: Toft, S. & Riedel, W. (Eds.), *Arthropod natural enemies in arable land. I. Density, spatial heterogeneity and dispersal*, Acta Jutlandica, 70 Aarhus University Press, Denmark, pp. 283-293.

Forman, R.T.T. & Baudry, J., 1984. Hedgerows and hedgerow networks in landscape ecology. *Environmental Management* **8**: 495-510.

Frampton, G.K., Cilgi, T., Fry, G.L.A. & Wratten, S.D., 1995. Effects of grassy banks on the dispersal of some carabid beetles (Coleoptera: Carabidae) on farmland. *Biological Conservation* **71**: 347-355.

Fry, G.L.A., 1991. Conservation in agricultural ecosystems. In: Spellerberg, I.F., Goldsmith, F.B. & Morris, M.G. (Eds.), *The scientific management of temperate communities for conservation*, Blackwell, Oxford, pp. 415-443.

Good, J. & Giller, P.S., 1991. The effect of cereal and grass management on staphylinid (Coleoptera) assemblages in southwest Ireland. *Journal of Applied Ecology* **28**:810-826.

Gourov, A.V., 1994. Territorial mosaic and the problem of boundaries (in case of succession). In: Cattaneo, D. & Semenzato, P. (Eds.), *Landscape ecology: Ecologia del Paesaggio*, Corso di Cultura in Ecologia, University of Padova, pp. 97-117.

Greaves, M.P. & Marshall, E.J.P., 1987. Field-margins: definitions and statistics. *British Crop Protection Council Monograph* **35**:1-10.

Gurr, G.M., Wratten, S.D. & van Emden, H.F., 1998. Habitat manipulation and natural enemy efficiency: implications for the control of pests. In: Barbosa, P. (Ed.), *Conservation Biological Control*. Academic Press, Cambridge, pp. 155-184.

Hagen, K.S., Viktorov, G.A., Yasumatsu, K. & Schuster, M.F., 1976. Biological control of pests of range forage and grain crops. In: Huffaker, C.B. & Messenger, P.S. (Eds.), *Theory and practice of biological control*, Academic Press, New York, pp. 397-442.

Hawthorne, A. & Hassell, M., 1995. The effect of cereal headland treatments on carabid communities. In: Toft, S. & Riedel, W. (Eds.), *Arthropod natural enemies in arable land. I. Density, spatial heterogeneity and dispersal*, Acta Jutlandica, 70 Aarhus University Press, Denmark, pp. 185-198.

Kaule, G. & Krebs, S., 1989. Creating new habitats in intensively used agricultural land. In: Buckley, G.P. (Ed.), *Biological habitat reconstruction*, Belhaven Press, London, pp. 161-169.

Kromp, B. & Nitzlader, M., 1995. Dispersal of ground beetles in a rye field in Vienna, Austria. In: Toft, S.

& Riedel, W. (Eds.), *Arthropod natural enemies in arable land. I. Density, spatial heterogeneity and dispersal*, Acta Jutlandica, 70 Aarhus University Press, Denmark, pp. 269-277.

Lageröf, J. & Wallin, H., 1993. The abundance of arthropods along two field margins with different types of vegetation composition - an experimental study. *Agriculture, Ecosystems and Environment* **43**:141-154.

Landis, D.A. & Haas, M.J., 1992. Influence of landscape structure on abundance and within field distribution of European corn borer (Lepidoptera: Pyralidae) larval parasitoids in Michigan. *Environmental Entomology* **21**:409-416.

Liebhold, A.M., Rossi, R.E. & Kemp, W.P., 1993. Geostatistics and Geographic Information Systems in applied insect ecology. *Annual Review of Entomology* **38**:303-327.

Lubbe, E., 1988. National report on environmental management in agriculture: W. Germany. In: Park, J.R. (Ed.), *Environmental management and agriculture, European perspectives*, Belhaven Press, London, pp. 83-94

Luff, M.L., 1966. The abundance and diversity of the beetle fauna of grass tussocks. *Journal of Animal Ecology* **35**:189-208.

Mader, H.J., 1988. Effects of increased spatial heterogeneity on the biocenosis in rural landscapes. *Ecological Bulletins* **39**:169-179.

Mauremootoo, J.R. & Wratten, S.D., 1994. Permeability of field boundaries to predatory carabid beetles. Proceedings IOBC/WPRS Working Group "Integrated Control in Cereals", *IOBC/WPRS Bulletin* **17**: 188-200.

Parish, T., Lakhani, K.H. & Sparks, T.H., 1994. Modelling the relationship between bird population variables and hedgerow and other field margin attributes. I. Species richness of winter, summer and breeding birds. *Journal of Applied Ecology* **31**:764-775.

Petit, S. & Burel, F., 1998. Effects of landscape dynamics on the metapopulations of a ground beetle (Col.: Carabidae) in a hedgerow network. *Agriculture, Ecosystems & Environment* **69**:243-252.

Pollard, E., Hooper, M.D. & Moore, N.W., 1974. *Hedges*. Collins, London,170 pp.

Potts, G.R., 1980. The effects of modern agriculture, nest predation and game management on the population ecology of partridges (*Perdix perdix* and *Alectoris rufa*). *Advances in Ecological Research* **11**:2-79.

Riedal, W., 1991. Overwintering and spring dispersal of *Bembidion lampros* Herbst. (Col.: Carabidae) from established hibernation sites in a winter wheat field in Denmark. In: Polgar, L., Chambers, R.J., Dixon, A.F.G. & Hodek, I. (Eds.), *Behaviour and impact of aphidophaga*, SBS Academic Publishing b.v., The Hague, pp. 235-241.

Riedel, W., 1995. Spatial distribution of hibernating polyphagous predators within field boundaries. In: Toft, S. & Riedel, W. (Eds.), *Arthropod natural enemies in arable land. I. Density, spatial heterogeneity and dispersal*, Acta Jutlandica, 70 Aarhus University Press, Denmark, pp. 283-293.

Rushton, S., Sanderson, R., Luff, M. & Fuller, R., 1996. Modelling the spatial dynamics of ground beetles (Carabidae) within landscapes. *Annales Zoologicae Fennici* **33**:233-241.

Saunders, D.A. & Hobbs, R.J., 1989. Corridors for conservation. *New Scientist* **63**:8.

Sherratt, T.N. & Jepson, P.C., 1993. A metapopulation approach to modelling the long term impacts of pesticides on invertebrates. *Journal of Applied Ecology* **30**:696-705.

Sotherton, N.W., 1985. The distribution and abundance of predatory Coleoptera overwintering in field boundaries. *Annals of applied biology* **106**:17-21.

Sotherton, N.W., 1991. Conservation headlands: a practical combination of integrated cereal farming and conservation. In: Firbank, L.G., Carter, N., Derbyshire, J.F. & Potts, G.R. (Eds.), *The ecology of temperate cereal fields*, Blackwell Scientific Publications, Oxford, pp. 193-197.

Sunderland, K.D., Crook, N.E., Stacey, D.L. & Fuller, B.J., 1987. A study of feeding by polyphagous predators on cereal aphids using ELISA and gut dissection. *Journal of Applied Ecology* **24**:907-933.

Thomas, M.B. & Wratten, S.D., 1988. Manipulating the arable environment to enhance the activity of predatory insects. *Aspects of Applied Biology* **17**:57-66.

Thomas, M.B., Wratten, S.D. & Sotherton, N.W., 1991. Creation of island habitats in farmland to manipulate populations of beneficial arthropods: predator densities and emigration. *Journal of Applied Ecology* **28**:906-917.

Thomas, M.B., Mitchell, H.J. & Wratten, S.D., 1992. Abiotic and biotic factors influencing the winter distribution of predatory insects. *Oecologia* **89**:78-84.

van Emden, H.F., 1965. The role of uncultivated land in the biology of crop pests and beneficial insects. *Scientific Horticulture* **17**:121-136.

Wallin, H., 1986. Habitat choice of some field inhabiting carabid beetles (Coleoptera: Carabidae) studied by recapture of marked individuals. *Ecological Entomology* **11**:457-466.

Wratten, S.D. & Thomas, C.F.G., 1990. Farm-scale spatial dynamics of predators and parasitoids in agricultural landscapes. In: Bunce, R.G.H. & Howard, D.C. (Eds.), *Species dispersal in agricultural landscapes*, Belhaven Press, London, pp. 219-237.

CHAPTER 13

NATURAL VEGETATION IN AGROECOSYSTEMS
Pattern and Scale of Heterogeneity

JOHN E. BANKS
Interdisciplinary Arts and Sciences, University of Washington, Tacoma, Tacoma, WA

13.1 Introduction

As today's natural world becomes increasingly fragmented, formerly continuous habitats begin to resemble checkerboards composed of disparate pieces of various shapes and sizes. The ensuing landscape structure has considerable consequences for both managed and natural populations throughout the world. Even as we become more aware of the negative effects of fragmentation on native communities, we seek to turn the tables in managed habitats by using habitat heterogeneity to our advantage in the quest to keep agricultural pest populations in check. As conservationists and agricultural researchers alike share growing concern over the fates of populations they try to preserve or eradicate, respectively, it is becoming increasingly clear that landscape fragmentation is no longer simply a concern for environmental activists.

An ecological perspective of the effects of habitat heterogeneity promises to be a powerful tool for managing agricultural systems. As concerns about the detrimental environmental effects of chemical pest control grow and traditional pesticides lose effectiveness or are outright banned in some countries, there is increasing pressure to develop safer and more sustainable means of pest control. Recently, much research has focused on determining the potential for controlling pest populations by manipulating vegetation diversity in agroecosystems. In particular, much effort has been put into the effects of mixed planting schemes in which primary crop plants are embedded in a matrix of other plants, which may be other crop plants or simply natural, non-crop vegetation. Two and half decades of such studies has left a legacy of species- and system- specific results (see Risch *et al.*, 1983; Andow, 1991; Tonhasca & Byrne, 1994 for reviews). The current challenge is to develop a more general theory that could be implemented in a variety of agricultural settings.

In this chapter, I discuss results of manipulative experiments designed to elucidate insect responses to different aspects of habitat heterogeneity. Using mixtures of a crop plant (broccoli) and natural weedy vegetation, I explored the effects of heterogeneity and the scale at which that heterogeneity was deployed on insect herbivores and predators. While these explorations originated as a means of learning something about the practical aspects of incorporating natural vegetation into crop fields as a technique for reducing reliance upon chemical control of pests, they ultimately yielded insight into more general

B. Ekbom, M. Irwin and Y. Robert (eds.), Interchanges of Insects, 215-229

issues concerning the role of scale in ecological experiments. Both perspectives should prove useful to researchers and growers interested in the possibility of developing a general theory of the effects of heterogeneity on insect populations in agroecosystems.

13.2 Mechanisms Underlying Insect Response to Heterogeneity

A few proposed mechanisms governing insect population responses to habitat diversification have emerged as theories that have been tested extensively in the field. The first mechanism, the "resource concentration hypothesis" stipulates that more diverse habitats should have lower pest densities because increased diversity inhibits the ability of herbivores to find host plants and increases their tendency to emigrate from host plants (Root, 1973). The second theory, the "natural enemies hypothesis", conjectures that more diverse habitats tend to harbor more predators and hence offer more protection against herbivores (Root, 1973).

These two proposed mechanisms are by no means mutually exclusive (Russell, 1989), and in many cases probably elements of both are in effect. Decades of tests of these mechanisms have generated equivocal results about the effects of habitat heterogeneity in general. A recent meta-analysis indicated that diversification has a low to moderate effect on herbivores (Tonhasca & Byrne, 1994). Another review of one hundred and fifty habitat diversification studies indicated that a simple majority of pest population densities were lower in more diversified habitats than in monocultures (Risch *et al.*, 1983). However, more than 35% of the studies in the review resulted in unchanged or even increased populations with increasing habitat complexity (Risch *et al.*, 1983; Andow, 1991). This ambiguity is not that surprising given that the studies encompassed a variety of insects with diverse life histories and movement behaviors; any general theory regarding the usefulness of habitat heterogeneity clearly will have to account for differences among insects being studied. At this point, the challenge is to determine which differences are important.

One notable feature of such studies, as well as of agriculture itself, is the scale at which diversification is deployed. Temperate farming schemes (*e.g.* in the U.S. and Europe) typically span several orders of magnitude in size, ranging from small organic farms of only a few hectares to intermediate-sized family farms tens of hectares in size to enormous agribusiness monocultures covering hundreds and thousands of hectares. The importance of scale, which has recently been given increasing attention in ecological circles (Murphy, 1989; Wiens, 1989; Wiens & Milne, 1989; Rose & Leggett, 1990; Doak *et al.*, 1992; Levin, 1992; Molumby, 1995), may find its most practical application in agroecosystems. While many experiments have explored the impacts of vegetation patterning on phytophagous insects (Pimental, 1960; Tahvanaienen & Root, 1972; Root, 1973; Cromartie, 1975; Andow, 1983; Bergelson & Kareiva, 1983; Letourneau & Altieri, 1983; Letourneau, 1987; Bach, 1980a,b, 1986, 1988; Elmstrom *et al.*, 1988; Lawrence & Bach, 1989; Perfecto, 1992; Luther *et al.*, 1996), few have explicitly addressed the consequences of the scale at which the patterning is expressed. As the scale of vegetation patterning directly impacts the ability of insects to move through and among habitats, it is likely to influence the distributions of pests in diverse habitats.

A common and widespread application of habitat diversification is the incorporation of natural vegetation in or around crop areas in an attempt to reduce pest populations. As this is simply a special case of habitat diversification discussed above, we can expect that the two proposed mechanisms (resource concentration and natural enemies hypotheses) might also be brought to bear on scenarios involving mixtures of a single crop and natural, weedy vegetation. One well-known example of this technique, also known as "weedy culture", is the hedgerow of British garden fame. Aerodynamic effects notwithstanding, the anticipated benefits of this type of habitat manipulation are reduced colonization of crop plants by herbivores, as well as increased emigration from host plants, and more abundant natural enemies.

In the United States, organic growers regularly grapple with decisions of weed management. Prohibited from using synthetic chemical herbicides, they must decide whether to pay exorbitant costs for the labor-intensive job of keeping their farms weed-free, or to leave some or all weeds in their crop areas. As a result, there is pronounced interest in the potential benefits of incorporating weeds into crop planting schemes.

Traditionally, studies of systems in which natural vegetation has been incorporated have principally focused on enhancing natural enemy populations as a means of reducing pest populations (van Emden, 1962; Galecka, 1966; Dempster, 1969; Flaherty, 1969; Perrin, 1975; Topham and Beardsley, 1975). However, in addition to providing more habitat suitable for a wider variety of predators, increasing natural vegetation in and around crop fields alters the ability of insects to move through the habitat. Ecologists have long known that dispersal ability can greatly influence the persistence and spatial distribution of populations, including both predators and prey (Gause, 1927; Huffaker, 1958; Luckinbill, 1973; Kareiva, 1987; Kruess & Tscharntke, 1994; Banks, 1997; Zabel & Tscharntke, 1998). In the experiments I describe here, I examine both herbivore and predator responses to habitat manipulations.

13.3 Experimental System and Design

In order to learn more about within-crop-field heterogeneity with respect to both colonizing herbivores as well as predators, I sampled species from both trophic levels and analyzed their distributions in response to habitat manipulations. By varying two factors of heterogeneity, I sought to answer two questions: (i) do different insects respond individualistically to different aspects of heterogeneity? and (ii) do experimental results vary as a function of the scale at which they are performed? The answers to these questions have consequences for both the way we do agriculture as well as the way we interpret past, present and future agroecological experiments.

I chose for my experimental system the common crop broccoli (*Brassica oleracea*) because it has a suite of specialized insects that feed on it in the Pacific Northwest of the U.S., where I conducted my experiments at Farm Two of Washington State University's Puyallup Research and Extension Center. My experimental design consisted of long narrow arrays, 2m wide and 32m long, in which patches of broccoli and patches of naturally occurring vegetation were grown in the various combinations as shown in Figure 1. The overall design was a two-way factorial analysis of variance in which three levels of heterogeneity (25%, 50%, and 75% broccoli monoculture) were crossed with three different levels of scale (4 m, 8 m, and 16 m wavelengths). The scale factor was indexed by the total length of a repeating unit of one broccoli patch and one weed patch ("wavelength"). Each array, then, represented one treatment combination of spatial heterogeneity (composition) and scale. The total experiment consisted of three replicates of each treatment array, with arrangement of arrays randomized within replicates. Areas both between arrays (at least 3 m) and within arrays in crop plant areas were kept bare through repeated cultivating and hoeing.

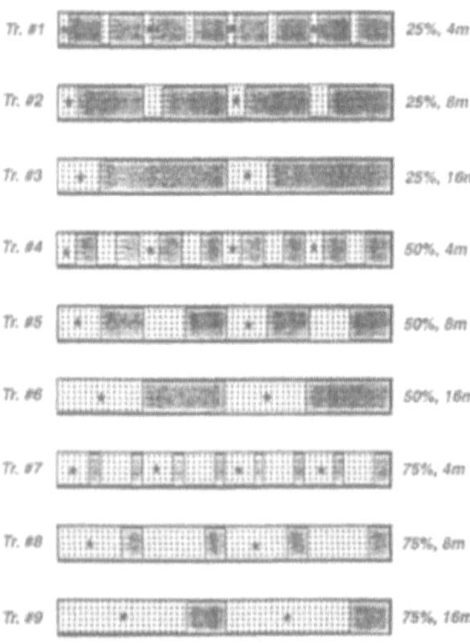

Figure 1. Schematic of all nine treatment arrays. Dotted areas represent broccoli patches, shaded areas denote weeds. Numbers to the right of each strip indicate percent of strip consisting of broccoli monoculture, and wavelength, respectively. Asterisks indicate patches in which broccoli patches were visually censused.

In non-broccoli areas, weeds were allowed to colonize and to grow naturally. This weedy vegetation consisted principally of *Echinochloa colonum* (L.), *Echinochloa crusgalli* (L.), *Cirsium arvense* (L.), *Chenopodium album* (L.), *Polygonum lapathifolium* (L.), *Amaranthus powellii*, *Equisetum arvense* (L.), *Sonchus asper* (L.), and *Lamium amplexicaule* (L.). I ran the experiment for two growing seasons, visually censusing herbivores on plants on average every ten days in the, 1994 and 1995 growing seasons. I also sampled insect predators during both seasons, either visually or via pitfall cup traps, albeit at larger time intervals. I performed a multivariate analysis of variance on densities of each insect species separately to determine treatment effects on populations, with the different census dates as my multiple variables (Wilkinson, 1992; Scheiner, 1993; von Ende, 1993).

I recorded densities for the three dominant herbivores in the system: *Brevicoryne brassicae* (cabbage aphids), *Phyllotreta cruciferae* (flea beetles), and eggs and larvae of *Pieris rapae* (the cabbage white butterfly). These three crucifer specialists were chosen because they are by far the most common herbivores in the system; in addition, they represent extremes in a gradient of dispersal ability and foraging behavior. The most prevalent predators were carabid beetles (*Pterostichus melanarius*) and ladybird beetles (*Coccinella septempunctata*); these also have contrasting movement/behavior characteristics. I discuss herbivore and predator results separately, after outlining behavioral traits of the study insects; more details may be found elsewhere (Banks, 1998, 1999).

13.4 Behavioral / Life History Characteristics of Study Herbivores

B. brassicae (Homoptera: Aphididae) alatae colonize crucifer host plants in early summer, typically settling down once a suitable plant is found and parthenogenically giving rise to apterous offspring. A single mature broccoli plant may harbor several thousand of these phloem-feeding homopterans. When conditions are right and a critical density of aphids is reached, production of new alatae is triggered; these new individuals then take off and fly to a new suitable host plant (Dixon, 1973). Apterous individuals, especially fourth-instar and young adults with high reproductive potential, are also capable of dispersing (Hodgson, 1991). Throughout most of the growing season, then, most local dispersal is performed by apterous individuals moving slowly from plant to plant, with occasional longer trips made by new colonizing alatae.

P. cruciferae (Coleoptera: Chrysomelidae) larvae pupate in the soil and colonize broccoli plants as adults. Although these chrysomelids can fly when colonizing host plants, their within-field movement is primarily characterized by a "flea-like" hop which propels them on average 25cm per jump (Vincent & Stewart, 1983). They are adept at adjusting their dispersal behavior in accordance with the distribution of host plants (Kareiva, 1982; Elmstrom *et al.*, 1988). While they have the capacity to move great distances, they may spend up to a day or two on or around a suitable host plant, chewing characteristically large pits in leaves.

P. rapae (Lepidoptera: Pieridae) adult females oviposit on cruciferous vegetables, laying eggs singly, usually on the underside of leaves. Adults tend to follow linear flight paths until they find themselves in the vicinity of host plants, when they switch to a more concentrated "oviposition" flight mode and skillfully hone in and oviposit on suitable crucifers (Jones *et al.*, 1980; Kareiva & Shigesada, 1983; Root & Kareiva, 1984). Larvae tend to stay on their natal plant, consuming large portions of leaves and only rarely moving to nearby host plants. They graze plants throughout their larval development period.

13.5 Behavioral / Life History Characteristics of Study Predators

P. melanarius is a generalist predator common in agricultural areas. It is active primarily at night, spending most of its time scurrying about the ground feeding on aphids, lepidopteran larvae, spiders, other beetles, and other insects. Since most individuals are micropterous during the growing season, they are easily captured in pitfall cup traps; this standard technique measures beetle activity (rather than actual distribution) (Lovei & Sunderland, 1996), but still allows for a meaningful comparison among the different habitat treatments.

C. septempunctata is a voracious predator, feeding primarily on aphids. A European native, it was introduced into the United States in the late 1950's as a potential biological control agent (Elliott *et al.*, 1996); by the 1970's it was well established in the eastern U.S., spreading to the west coast by the 1980's. These ladybird beetles fly readily among patches of plants, then typically walk from branch to branch of individual plants in search of aphid colonies.

13.6 Results of Vegetation Manipulations: Herbivores

The multivariate analysis of variance, for which herbivore densities for each census across two growing seasons served as response variables, revealed that the three focal herbivores responded individualistically to different aspects of heterogeneity (Table 1). Cabbage aphids responded strongly to simple heterogeneity (% crop cover), but not to scale (wavelength). In contrast, flea beetles exhibited a significant response to scale manipulations, as well as to an interaction between simple heterogeneity and scale. Finally, cabbage white butterfly densities were unaffected by heterogeneity patterning at all scales.

Table 1. Results of MANOVA on herbivore densities in response to two heterogeneity treatment effects (scale and percent crop cover) across two growing seasons ($p < 0.05$). See Banks (1998) for details

SPECIES	DISPERSAL ABILITY (COLONIZATION / LOCAL MOVEMENT)	TREATMENT EFFECTS?
CABBAGE WHITE	EXCELLENT / POOR	NONE
FLEA BEETLES	FAIR / INTERMEDIATE	SCALE; INTERACTION
APHIDS	FAIR / POOR	% CROP COVER

While the MANOVA serves to identify overall differences in herbivore responses to the field manipulations, it indicates nothing about the direction of those differences. Furthermore, it is problematic to graphically display the results of a MANOVA, since it synthesizes many univariate responses and is not amenable to being portrayed as a single graph. Instead, I present here herbivore densities from a mid-season census for the 1994 growing season; I present these particular results because they illustrate the general trend in herbivore densities across all of the censuses (see Banks, 1998 for more details).

Cabbage aphids attained higher numbers in more weedy treatments than in arrays consisting of mostly crop plants (Fig. 2). While the particular quantitative difference varied among census dates, the qualitative relationship portrayed in Figure 2 illustrates a trend across both growing seasons. This result does not support traditional theory concerning the effects of heterogeneity on herbivore densities, which predicts higher herbivore densities as monoculture area increases (Root, 1973). One possible explanation is that, given the propensity of aphids to perceive and drop into crop plant patches in response to light reflection intensity (Kring, 1972), aphids in the air column over 25% crop cover treatment arrays funneled into and accumulated in the smaller patches in those arrays. Since their subsequent movement after colonization is limited (as the population consists mostly of apterous individuals), aphid distributions are probably largely determined by the colonization behavior of alates arriving early in the season.

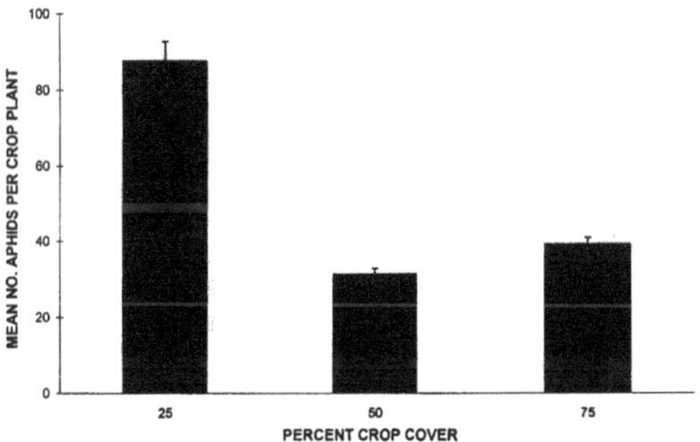

Figure 2. Mean number of aphids per plant as a function of percent crop cover averaged over all wavelengths for all treatment arrays in a mid-season census, July 1994. Bars represent SE for three means (from three replicates) per wavelength (N = 9).

Flea beetles exhibited a more complex response to the habitat manipulations. While the MANOVA detected an overall significant response to the scale treatment effect, it also indicated that flea beetles responded to an interaction between scale (wavelength) and simple heterogeneity (*i.e.* percent crop cover). One way of interpreting this is that the impact of percent crop cover on flea beetles varied with the scale at which the heterogeneity was presented. A closer look at flea beetle distributions from a mid-season census in 1994 (Fig. 3) illustrates this interaction: the difference between flea beetle densities at the intermediate percent crop cover and densities at the other percent crop covers was much greater at the largest (16 m) wavelength than at the smaller scales. Furthermore, flea beetles attained either their highest or lowest densities at intermediate crop cover treatments, depending on the scale of the patterning manipulation. These results highlight the importance of performing experimental manipulations at a variety of scales when trying to assess the influence of heterogeneity on different organisms. For instance, if this experiment had been conducted only at the smallest or largest scale, we might have concluded that intermediate crop cover levels consistently result in the highest flea beetle densities. Similarly, we might note that although the cabbage white butterfly responded to none of the mixture treatments, it's possible that they would respond to habitat manipulations at scales larger than those presented in this experiment.

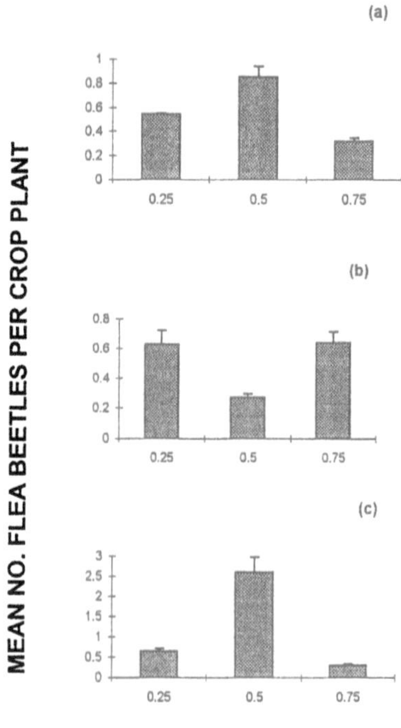

Figure 3. Mean number of flea beetles per plant as a function of percent crop cover broken down by wavelength (4m for (a), 8m for (b), and 16m for (c)) in a mid-season census, July 1994. Bars represent SE for three means (from three replicates) (N = 3).

13.7 Results of Vegetation Manipulations: Predators

The MANOVA for carabid beetle and ladybird beetle densities revealed that while overall carabids did not respond to either type of heterogeneity manipulation, ladybird beetles responded to the scale at which patterning was presented in the field (Table 2).

Table 2. Results of MANOVA on predator densities in response to two heterogeneity treatment effects (scale and percent crop cover) across two growing seasons ($p < 0.05$). See Banks (1999) for details

SPECIES	DISPERSAL CHARACTERISTICS	TREATMENT EFFECTS?
CARABIDS	GROUND DWELLING	NONE
LADYBIRD BEETLES	FLY TO PLANTS, WALK FROM PLANT TO PLANT	SCALE

A look at typical coccinellid densities as a function of scale reveals that ladybird beetles were on average much more abundant in arrays with larger wavelengths (Fig. 4).

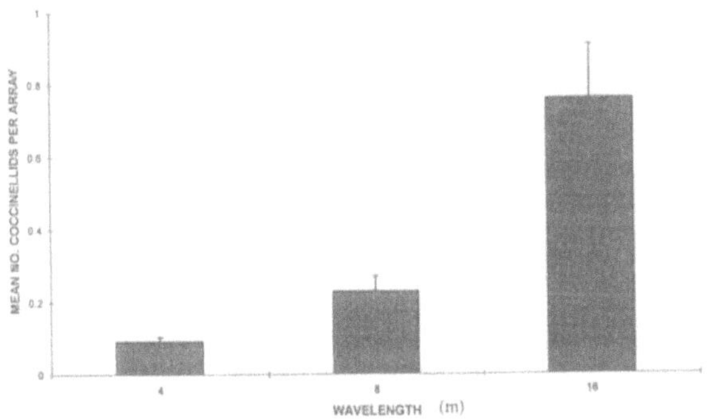

Figure 4. Mean number of coccinellids per arrays as a function of wavelength (m) averaged over all crop covers in a mid-season census, July 1994. Bars represent SE for three means (from three replicates) per percent crop cover treatment (N = 9).

The effects of habitat patchiness on coccinellids are well documented: increasing patchiness has been shown to diminish ladybird beetles' ability to aggregate to prey and thus increase their tendency to emigrate (Kareiva, 1987). Since arrays with smaller wavelengths necessarily have increased amounts of patchiness, it seems as though a similar mechanism may be at work in this system. If we delve further into the details of precisely where coccinellids are found within treatment arrays, we find that coccinellids were disproportionately concentrated in crop areas in the least weedy arrays (Fig. 5). That is, if ladybird beetles were raining down into treatment arrays with no regard for habitat type, we would expect that 25% of beetles found in 25% crop cover treatments would be found in crop areas, 50% of beetles found in 50% crop cover would be found in crop areas, and so on - falling on the diagonal line in Figure 5. Instead, nearly 90% of beetles in 75% crop cover arrays were found in crop areas (Fig. 5), suggesting that coccinellids aggregate more to crop areas (and to prey areas) in less weedy treatments.

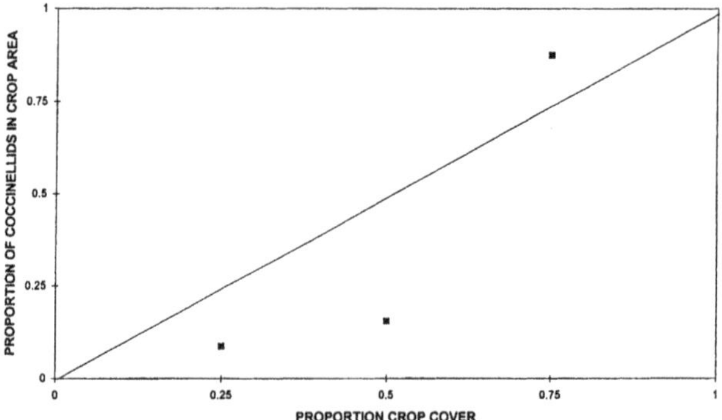

Figure 5. Mean proportion of coccinellids found in crop areas within treatment arrays, averaged across all wavelengths, as a function of percent crop cover. Data are from a mid-season census in July 1994. Straight 45-degree line represents expected values if coccinellids colonized treatment arrays with no preference for weed or crop habitat.

While the MANOVA detected no significant response of carabid beetles to habitat manipulations, it is worth taking a closer look at the details of their distributions just as we did for the coccinellids, since the distribution of beetles within arrays is arguably an important indicator of their efficacy as biological control. If we look at the proportion of beetles actually found within crop areas for the different treatment arrays, we see that beetles were slightly overrepresented in crop areas in the weediest arrays (Fig. 6). That is, nearly 30% of beetles were found in crop areas in arrays consisting of 25% crop cover. As carabids are known to have a penchant for weedy ground cover (Lovei & Sunderland, 1996), it is possible that they feel more comfortable venturing out into crop areas in more weedy areas, where they are always on average closer to weed cover.

We need more detailed behavioral observations if we are to make quantitative predictions about how best to manipulate habitat in order to foster these predator populations.

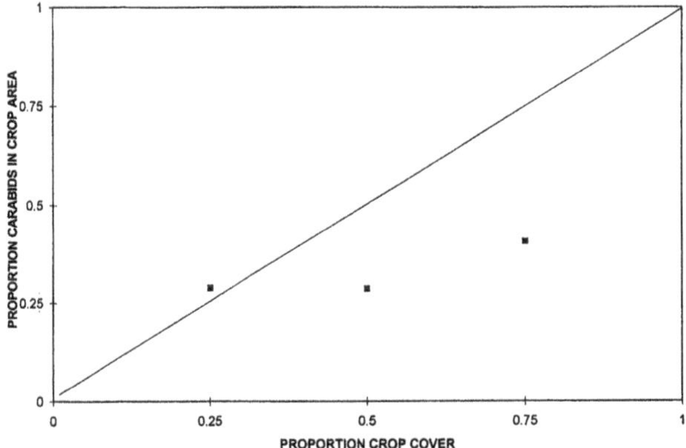

Figure 6. Mean proportion of carabids found in crop areas within treatment arrays, averaged across all wavelengths, as a function of percent crop cover. Data are from a mid-season census in July 1994. Straight 45-degree line represents expected values if carabids colonized treatment arrays with no preference for weed or crop habitat.

It is clear, however, from these results, that rather than looking for a generic pattern of habitat heterogeneity, we would do well to consider the individualistic responses of different predators. For instance, if we were to apply the results presented here into a recipe for enhancing both carabid beetle and coccinellid populations, we might prescribe deploying crops in (i) weedy and (ii) large continuous patches. Of course, the exact quantitative specifications will depend on movement rates, reproduction, etc., and will be subject to abiotic constraints (*e.g.* economic factors, yield considerations, etc.), which are not considered in this simplistic speculation. In general, however, the qualitative solutions for such a complex problem will most likely consist of an optimization procedure into which we incorporate parameters from the entire community rather than single-species responses to heterogeneity.

13.8 What Have We Learned?

The results described here are simply a start towards elucidating the response of insect populations to landscape heterogeneity. While they do not yet present a general quantitative recipe for deploying habitat heterogeneity, they do point out several important subtleties that we might keep in mind in the future. First, although it is clear from previous studies that different insects respond individualistically to heterogeneity, these results emphasize that insects respond differently even to different *aspects* of heterogeneity. That is, we cannot expect to find a general heterogeneity template that will reduce pest populations or enhance predator populations across the board. Rather, we would do well to incorporate individual species' responses to various forms of heterogeneity into a type of optimization algorithm, so that we might maximize pest reduction or predator persistence, effecting a compromise that results in the best overall pest reduction. Naturally, this optimization process would necessarily include an assessment of current economic damage attributable to different pests. Although this is logistically problematic for long-term strategies such as diversity manipulations, it is an important reminder of the need for a more economics-based paradigm in agroecology.

These results also provide a cautionary tale for those interested in extrapolating results from small ecological experiments to larger systems. The fact that some insects' (*e.g.* flea beetles) responses to simple heterogeneity are scale-dependent serves as a warning to those who would (i) perform future ecological experiments at one scale only, and (ii) assume that results from previous work done at a single scale are invariant across scales. It is clear that we have much to learn by explicitly incorporating scale into experimental designs, especially if we are doing agroecology experiments in which "scaling up" is inevitable.

13.9 Future Directions

While the results of these experiments are a first step towards developing a protocol for deploying heterogeneity as a means of pest control, we certainly need more detailed studies. Only by performing thorough behavioral observations of pests and predators data can we expect to flesh out mechanisms underlying distributional data such as those presented here. Furthermore, due to inevitable fluctuations in population dynamics, field conditions, etc. throughout the growing season, it is unlikely that we will ever be able to rely on habitat heterogeneity as the sole means of pest control. However, there is hope for incorporating heterogeneity into an effective integrated pest management (IPM) regimen. As we learn more about the interaction of natural enemies and pesticides (*e.g.* Gould, 1991; Johnson & Gould, 1992; Yardim & Edwards, 1998), the use of natural vegetation in agroecosystems may emerge as an important factor in controlling herbivore populations (Banks & Stark, 1998). With a full understanding of how and why heterogeneity acts to influence pest and predator populations, we should be able to reduce our reliance on synthetic pesticides. Such a reduction would be welcome to both conservationists and growers alike in the face of ongoing ecotoxicological mishaps and ever-increasing restrictions on chemical use in agriculture (Banks & Stark, 1998).

Acknowledgments

Thanks to P. Kareiva, R.T. Paine, D. Schemske, and R. Huey for helpful comments and editing. I am grateful to B. Fagan, A. Kruess, T. Tscharntke, B. Ekbom, and R. Bommarco for generously contributing ideas and thoughtful discussions. Special thanks to J. Stark for providing land and facilities. This work was supported by a Sigma Xi Grant-in-Aid-of-Research, and a National Science Foundation Dissertation Improvement grant (DDR 61-4582).

13.10 References

Andow, D., 1983. The extent of monoculture and its effects on insect pest populations with particular reference to wheat and cotton. *Agric. Ecosyst. Envt.* **9**: 25-35.

Andow, D., 1991. Vegetational diversity and arthropod population response. *Ann. Rev. Entomol.* 36: 561-586.

Bach, C.E., 1980*a*. Effects of plant density and diversity on the population dynamics of a specialist herbivore, the striped cucumber beetle, *Acalymma vittata* (Fab). *Ecology* **61**: 1515-1530.

Bach, C.E., 1980*b*. Effects of plant diversity and time of colonization on an herbivore-plant interaction. *Oecologia* **44**: 319-326.

Bach, C.E., 1986. A comparison of the responses of two tropical specialist herbivores to host plant size. *Oecologia* **68**: 580-584.

Bach, C.E., 1988. Effects of host plant patch size on herbivore density: patterns. *Ecology* **69**: 1090-1102.

Banks, J.E., 1997. Do imperfect tradeoffs affect the extinction debt phenomenon? *Ecology* **78**: 1597-1601.

Banks, J.E., 1998. The scale of landscape fragmentation affects herbivore response to vegetation heterogeneity. *Oecologia* **117**: 239-246.

Banks, J.E., 1999. Differential response of two agroecosystem predators, *Pterostichus melanarius* (Coleoptera: Carabidae) and *Coccinella septempunctata* (Coloeptera: Coccinellidae), to habitat-composition and fragmentation-scale manipulations. *The Canadian Entomologist* (In Press)."

Banks, J.E. and J.D. Stark., 1998. What is ecotoxicology? An ad-hoc grab or an interdisciplinary science? *Integrative Biology* **1(5)**: 195-204.

Bergelson, J. & P. Kareiva., 1987. Barriers to movement and the response of herbivores to alternative cropping patterns. *Oecologia* **71**: 457-460.

Corbett, A. & R.E. Plant., 1993. Role of movement in the response of natural enemies to agroecosystem diversification: a theoretical evaluation. *Environ. Entomol.* **22**: 519-531.

Cromartie, W.J., 1975. The effect of stand size and vegetational background on the colonization of cruciferous plants by herbivorous insects. *J. Appl. Ecol.* **12**: 517-533.

Dempster, J.P., 1969. Some effects of weed control on the numbers of the small cabbage white (*Pieris rapae* L.) on brussels sprouts. *J. Appl. Ecol.* **6**: 339-345.

Dixon, A.F.G., 1973. The biology of aphids. Arnold, London.

Doak, D.F., Marino, P.C. & Kareiva; P.M., 1992. Spatial scale mediates the influence of habitat fragmentation on dispersal success: implications for conservation. *Theor. Pop. Biol.* **3**: 315-336.

Elliott, N., Kieckhefer, R. & Kauffman, W., 1996. Effects of an invading coccinellid on native coccinellids in an agricultural landscape. *Oecologia* **105**: 537-544.

Elmstrom, K.M., Andow, D.A. & Barclay, W.W., 1988. Flea beetle movement in a broccoli monoculture and diculture. *Environ. Entomol.* **17**: 299-305.

Flaherty, D.L., 1969. Ecosystem trophic complexity and densities of the Willamette mite, *Eotetranychus willamettei* Ewing (Acarina: Tetranychidae). *Ecology* **50**: 911-916.

Galecka, B., 1966. The effectiveness of predators in control of *Aphid nasturtii* Kalt. and *Aphis frangulae* Kalt. on potatoes, In: I. Hodek (Ed.), *Ecology of aphidophagous insects.* Academia, Prague, Czechoslovakia, pp. 255-258.

Gause, G.F., 1934. *The struggle for existence.* Macmillan (Hafner Press), New York, NY.

Gould, F., 1991. Arthropod behavior and the efficacy of plant protectants. *Ann. Rev. Entomol.* **36**: 305-330.

Hodgson, C., 1991. Dispersal of apterous aphids (Homoptera: Aphididae) from their host plant and its significance. *Bulletin of Entomological Research* **81**: 417-427.

Huffaker, C.B., 1958. Experimental studies on predation: Dispersion factors and predator-prey oscillations.

Hilgardia **27**: 343-383.

Jones, R., Gilbert, N. & V. Neals., 1980. Long-distance movement of *Pieris rapae*. *Journal of Animal Ecology* **49**:629-642.

Johnson, M.T. & F. Gould., 1992. Interaction of genetically engineered host plant resistance and natural enemies of *Heliothis virescens* (Lepidoptera: Noctuidae) in tobacco. *Environmental Entomology* **21**: 586-597.

Kareiva, P., 1982. Experimental and mathematical analyses of herbivore movement: quantifying the influence of plant spacing and quality on forage discrimination. *Ecological Monographs* **52**: 261-282.

Kareiva, P., 1987. Habitat fragmentation and the stability of predator-prey interactions. *Nature* **326**: 388-390.

Kareiva, P.M. & Shigesada; N., 1983. Analyzing insect movement as a correlated random walk. *Oecologia* **56**:234-238.

Kring, J.B., 1972. Flight behavior of aphids. *Annual Review of Entomology* **17**: 461-492.

Kruess, A. & Tscharntke, T., 1994. Habitat fragmentation, species loss and biological control. *Science* **264**: 1581-1584.

Kruess, A. & Tscharntke, T., 1999. Effects of habitat fragmentation on plant-insect communities. In: Ekbom, B., Irwin, M & Robert, Y. (Eds.), *Interchanges of insects between agricultural and surrounding landscapes*. Kluwer Academic Press, Dordrecht, The Netherlands.

Lawrence, W.S. and C.E. Bach., 1989. Chrysomelid beetle movements in relation to host-plant size and surrounding non-host vegetation. *Ecology* **70**: 1679-1690.

Letourneau, D., 1987. The enemies hypothesis: tritrophic interactions and vegetation diversity in tropical agroecosystems. *Ecology* **68**:1616-1622.

Letourneau, D. & Altieri, M.A., 1983. Abundance patterns of a predator, *Orius tristicolor* (Hemiptera: Anthocoridae), and its prey, *Frankliniella occidentalis* (Thysanoptera: Thripidae): habitat attraction in polycultures versus monocultures. *Environ. Entomol.* **12**: 1464-1469.

Levin, S.A., 1992. The problem of pattern and scale in ecology. *Ecology* **73**: 1943-1967.

Lovei, G.L. & Sunderland, K.D., 1996. Ecology and behavior of ground beetles (Coleoptera: Carabidae). Ann. *Rev. Entomol.* **41**: 231-256.

Luckinbill, L.S., 1973. Coexistence in laboratory populations of *Paramecium aurelia* and its predator *Didinium nasutum*. *Ecology* **54**: 1320-1327.

Luther, G.C., Valenzulea, H.R. & Defrank, J., 1996. Impact of cruciferous trap crops on lepidopteran pests of cabbage in Hawaii. *Environ. Entomol.* **25**: 39-47.

Molumby,A., 1995. Dynamics of parasitism in the organ-pipe wasp, *Trypoxylon politum*: effects of spatial scale on parasitoid functional response. *Ecol. Entomol.* **20**: 159-168.

Murphy, D.D., 1989. Conservation and confusion: wrong species, wrong scale, wrong conclusions. *Cons. Biol.* **3**: 82-84.

Perfecto, I. & Sediles, A., 1992. Vegetational diversity, ants, and herbivorous pests in a neotropical agroecosystem. *Environ. Entomol.* **21**:61-67.

Perrin, R.M., 1975. The role of perrenial stinging nettle, *Urtica dioica*, as a reservoir of beneficial insects. *Ann. Appl. Biol.* **81**: 289-297.

Pimental, D., 1960. Species diversity and insect population outbreaks. *Ann. Ent. Soc. Amer.* **54**: 76-86.

Root, R.B. & Kareiva, P.M., 1984. The search for resources by cabbage butterflies (*Pieris rapae*): ecological consequences and adaptive significance of Markovian movements in a patchy environment. *Ecology* **65**:147-165.

Rose,G.A. & Leggett,W.C., 1990. The importance of scale to predator-prey spatial correlations: an example of atlantic fishes. *Ecology* **71**:33-43.

Risch, S. J., Andow D. & Altieri, M.A., 1983. Agroecosystem diversity and pest control: data, tentative conclusions, and new research directions. *Environ. Entomol.* **12**: 625-629.

Scheiner, S.M., 1993. MANOVA: Multiple response variables and multispecies interactions. In: Scheiner, S.M. & Gurevitch, J. (Eds.), *Design and Analysis of Ecological Experiments*. Chapman and Hall, New York, NY, pp. 94-112

Tahvanainen, J.O. & Root, R.B., 1973. The influence of vegetational diversity on the population ecology of a specialized herbivore, *Phyllotreta cruciferae* (Coleoptera: Chrysomelidae). *Oecologia* **10**: 321-346.

Tonhasca, A. Jr. & Byrne, D.A., 1994. The effects of crop diversification on herbivorous insects: a meta-analysis approach. *Ecological Entomology* **19**: 239-244.

Topham, M. & Beardsley, Jr., J.W, 1975. Influence of nectar source plants on the New Guinea sugarcane weevil parasite, *Lixophaga spenophori* (Villeneuve). *Proc. Hawaii. Entomol. Soc.* **22**: 145-154.

van Emden, H.F., 1962. A preliminary study of insect numbers in field and hedgerow. *Entomol. Mon. Mag.* **98**: 255-259.

van Ende, C.N., 1993. Repeated-measures analysis: growth and other time-dependent measures. In: Scheiner, S.M. & Gurevitch, J. (Eds.), *Design and Analysis of Ecological Experiments*. Chapman and Hall, New York, NY, pp. 113-137.

Wiens, J.A., 1989. Spatial scaling in ecology. *Functional Ecology* **3**: 385-397.

Wiens, J.A. & Milne, B.T., 1989. Scaling of 'landscapes' in landscape ecology, or, landscape ecology from a beetle's perspective. *Landscape Ecology* **3**: 87-96.

Wilkinson, L., 1992. *SYSTAT: Statistics*, version 5.2. Systat Incorporated, Evanston, Illinois.

Yardim, E.N. & Edwards, C.A., 1998. The influence of chemical management of pests, diseases, and weeds on pest and predatory arthropods associated with tomatoes. *Agric. Ecosyst. Environ.* **70**: 31-48.

Zaber, J. & Tscharntke, T., 1998. Does fragmentation of *Urtica* habitats affect phytophagous and predatory insects differentially? *Oecologia* **116**: 419-425.

INDEX

GPSR Compliance
The European Union's (EU) General Product Safety Regulation (GPSR) is a set of rules that requires consumer products to be safe and our obligations to ensure this.

If you have any concerns about our products, you can contact us on

ProductSafety@springernature.com

In case Publisher is established outside the EU, the EU authorized representative is:

Springer Nature Customer Service Center GmbH
Europaplatz 3
69115 Heidelberg, Germany

www.ingramcontent.com/pod-product-compliance
Ingram Content Group UK Ltd.
Pitfield, Milton Keynes, MK11 3LW, UK
UKHW041444070726
13610UKWH00009B/13

* 9 7 8 9 4 0 1 7 1 9 1 4 8 *